"This book was originally published by the Food and Agriculture Organization of the United Nations (FAO) as Freshwater Prawn Farming - A Manual for the Culture of Macrobrachium Rosenbergii, FAO Fisheries Technical Paper 225 Rev. 1.

"The designations employed and the presentation of material in this publication do not imply the expression of any opinion whatsoever on the part of the Food and Agriculture Organization of the United Nations concerning the legal status of any country, territory, city or area or of its authorities, or concerning the delimitation of its frontiers or boundaries."

"The designations 'developed' and 'developing' economies are intended for statistical convenience and do not necessarily express a judgement about the stage reached by a particular country, territory or area in the development process."

First Indian Edition 1995

ISBN: 978-8-170-35151-1

Rs. 995

Published by : **Daya Publishing House**
1123/74 Deva Ram Park,
Tri Nagar, Delhi-110035
Phone : 7220593, 7231826

Printed at : **Replika Press Pvt. Ltd.**

PREPARATION OF THIS DOCUMENT

This is the revised version of FAO Fisheries Technical Paper No. 225, one of a series of documents periodically produced by the FAO Fishery Resources and Environment Division to help to meet the needs of aquaculture workers of Member Countries for synthesis of information in the field of aquaculture.

It was prepared under contract for FAO by the former Co-Managers of the UNDP/FAO Programme for the Expansion of Freshwater Prawn Farming in Thailand, Messrs. M.B. New and S. Singholka. If reflects mainly the Thai way of freshwater prawn farming, and tries to be as practical as possible, keeping in mind that simple methods are more rapidly applicable in developing countries.

It also includes many findings of the UNDP/FAO Programme for the Expansion of Freshwater Prawn Farming, which have contributed to the advancement of this aquaculture practice in Thailand.

Distribution:

FAO Fisheries Department
FAO Regional Fisheries Officers
Regional Projects
GFCM
CIFA
CECAF
IPFC
COPESCAL
Authors

For bibliographic purposes this document should be cited as follows:

New, M.B. and S. Singholka, Freshwater prawn
1985 farming. A manual for the culture of Macrobrachium rosenbergii. FAO Fish.Tech.Pap., (225)Rev.1: 118 p.

ABSTRACT

This manual has been prepared on the basis of Thai experience in the farming of the giant freshwater prawn *Macrobrachium rosenbergii*. It has a practical orientation, as it was written for extension and fishery officers as well as for practising aquaculturists willing to start freshwater prawn farming.

The farming of this prawn is of recent origin, but is spreading very rapidly in several countries which followed the well-divulgated example of Hawaiian prawn farmers. At the time in which this manual was written, Thailand had taken the world leadership in terms of production as the result of an active long-term involvement of both the official and private sectors.

Since there is already a great deal of mainly scientific literature on this species, this manual was conceived to fill a gap in the existing literature, in order to provide information on how to go about *M. rosenbergii* farming answering the numerous requests from Member Governments to FAO.

It contains some general information on the biology of the species and much more detailed information on larval rearing and pond culture. The Thai practice forms the core of the manual but reference to other alternative techniques, used elsewhere, is given as well as additional literature references relevent to each topic dealt with in the text, for readers willing to enlarge their knowledge.

Annexes on water filtration, nutrition of larvae, description of the various larval stages, pond feeds, seine set design, stock estimation, and management strategy for "continuous" culture technique, are also provided as well as a glossary of scientific and technical terms used in the main text.

CONTENTS

CONTENTS (continued)

CONTENTS (continued)

1. INTRODUCTION

This manual is intended as a practical guide to freshwater prawn farming primarily for extension, rather than research workers. Its contents are a synthesis of the practical experience of its two authors, blended with their observation of the work of others in this field.

Although several species of freshwater prawns are currently being cultured, this manual deals exclusively with the farming of the species *Macrobrachium rosenbergii*, which is indigenous to South and Southeast Asia, parts of Oceania and some Pacific islands. *M. rosenbergii* has been imported into many other tropical and subtropical areas of the world and is the species of the genus most favoured for farming purposes. The words, freshwater prawns, when used subsequently in this manual, refer specifically to *M. rosenbergii*.

The hatchery and pond-rearing techniques described are those generally applied to freshwater prawns in Thailand. Only one system of culture has been fully described, to simplify the manual, but reference is made to alternative commercial and experimental techniques and many references are provided for further reading.

It had originally been intended by the authors to write this manual in an abbreviated 'cookbook' form. However, there is so much art, as well as science, in successful freshwater prawn farming that this approach did not prove feasible. The manual has therefore been prepared in narrational form.

After a brief section on the biology of *M. rosenbergii*, sections 3-5 deal with hatchery siting, facilities and operation, while sections 6-8 deal with the siting, facilities and operation of ponds for rearing freshwater prawns to market size. Section 10 lists the papers and books referred to in the text. A number of important topics, such as the preparation of feed for freshwater prawn larvae, are covered in the Appendixes. A glossary of terms and conversions is given in Appendix 10.

The culture of other species of *Macrobrachium*, though by 1981 still on a limited scale, is taking place using modifications of the techniques described in this manual. These modifications take account of the different environmental requirements of other species, especially in the larval stages. Some reference to the culture of other *Macrobrachium* spp. is provided in the works of Goodwin and Hanson (1975) and Hanson and Goodwin (1977).

The authors of this manual hope that you will find it useful and stimulating. We would welcome constructive criticism so that the manual may be improved in future editions.

2. BIOLOGY

The following brief notes are provided to give some background information on the culture of freshwater prawns and to set the scene for this manual.

2.1 Distribution

Species of the freshwater prawn genus *Macrobrachium* are distributed throughout the tropical and subtropical zones of the world. Over 100 species are known to exist, over a quarter of these being found in the Americas. Holthuis (1980) has provided useful information on the distribution, local names, habitats and maximum sizes of commercial species of *Macrobrachium*.

They are found in most inland freshwater areas including lakes, rivers, swamps, irrigation ditches, canals and ponds, as well as in estuarine areas. Most species require brackish water in the initial stages of their life cycle (and therefore they are found in water that is directly or indirectly connected with the sea) although some complete their cycle in inland saline and freshwater lakes. Some species prefer clear-water rivers while others are found in extremely turbid conditions. *M. rosenbergii* is an example of the latter.

There is a wide interspecific variation in maximum size and growth rate, *M. rosenbergii*, *M. americanum* and *M. carcinus* being probably the largest species known. *M. americanum* is found naturally in western watersheds of the Americas while *M. carcinus* is found in those connected with the Atlantic. *M. rosenbergii* is indigenous in the whole of the South and Southeast Asian areas as well as in northern Oceania and in the western Pacific islands.

Many *Macrobrachium* species have been transferred from their natural location to other parts of the world, initially for research purposes. *M. rosenbergii* is the species most used for commercial farming and consequently is the one which has been introduced to more countries. Following its import into Hawaii from Malaysia in 1965-66, where the pioneer work of Ling (1969a) was translated into a method for the mass production of post-larvae by Fujimura and Okamoto (1972), it has been introduced into almost every continent for farming purposes. *M. rosenbergii* is now farmed in considerable quantity in many countries, including Hawaii, Honduras, Mauritius, Taiwan and Thailand; and farms are now being established in many others, including Costa Rica, Indonesia, Israel, Malaysia, Mexico, the Philippines and Zimbabwe.

2.2 Life History

In the context of freshwater prawn farming, the 'lay' terms 'shrimp' and 'prawn' are synonymous. Each is used depending on the country where the animal is known.

In order to grow, all freshwater prawns (like other crustaceans) have to regularly cast their 'exoskeleton' or shell. This process is referred to as moulting and is accompanied by a sudden increase in size and weight. There are four distinct phases in the life cycle of the freshwater prawn, namely egg, larva, post-larva and adult. The time spent by each species of *Macrobrachium* in the different phases of its life cycle and its growth rate and maximum size varies, not only specifically but according to environmental conditions, mainly temperature. The life cycle of *M. rosenbergii* can be summarized as follows:

Copulation of adults results in the deposition of semen in a gelatinous mass on the underside of the thoracic region of the female's body (between the walking legs). Successful mating can only take place between hard-shelled males and ripe females which have just completed their pre-mating moult and are soft-shelled. Under natural conditions, mating of *M. rosenbergii* occurs throughout the year, although there are sometimes peaks of activity related to environmental conditions. Egg-laying occurs within a few hours of copulation, the eggs being fertilized on extrusion by the semen attached to the exterior of the female's body. The eggs are transferred to a brood chamber of the underside of the abdominal region of the female, held in place by a thin membrane and kept aerated by vigorous movements of the abdominal appendages. The length of time that the eggs are carried by the female in this way varies but is not normally longer than three weeks. The number of eggs which are laid depends also on the size of the female. Female prawns of *M. rosenbergii* are reported to lay from 80 000 to 100 000 eggs during one spawning when fully mature. However, their first broods, which are produced within their first year of life, are often not more than 5 000 to 20 000. In laboratory conditions, where a breeding stock of both males and females was kept, it was noted that egg incubation time averaged 20 days at 28°C (range 18-23 days). Ovaries frequently ripened again while females were carrying eggs. Pre-mate intermoults were separated by as little as 23 days (i.e., females on some occasions hatched two batches of eggs within a one-month period. It is unlikely that this would happen under natural conditions but it does show the potential fecundity of the animal).

As the eggs hatch, a process which is normally completed for the whole brood within one or two nights, the larvae are dispersed by rapid movements of the abdominal appendages of the parent. Freshwater prawn larvae are planktonic and swim actively tail first, ventral side uppermost. They require brackish water for survival. Those which hatch in fresh water will die unless they reach brackish water within a few days. There are a number of microscopically distinct stages during the larval life of freshwater prawns which lasts several weeks. Individual larvae of *M. rosenbergii* have been observed, in hatchery conditions, to complete their larval life in as little as 16 days. Larvae eat continuously and, in nature, their diet is principally zooplankton (mainly minute crustaceans), very small worms and the larval stages of other aquatic invertebrates.

On completion of their larval life, freshwater prawns metamorphose into post-larvae. From this point onwards they resemble miniature adult prawns and become mainly crawling rather than free swimming animals. When they do swim it is normally in a dorsal side uppermost fashion and in a forward direction. Rapid evasive movement is also achieved by contracting the abdominal muscles. Post-larvae exhibit good tolerance to a wide range of salinities which is a characteristic of freshwater prawns: newly metamorphosed hatchery-reared post-larvae can be rapidly transferred from brackish water to completely fresh water.

Post-larvae begin to migrate upstream into freshwater conditions within one or two weeks after metamorphosis and are soon able to swim against rapidly flowing currents or to crawl over the stones at the shallow edges of rivers and in rapids. They can climb vertical surfaces and cross land, provided there is abundant moisture available. In addition to using the foods available to them as larvae, they now utilize larger pieces of organic material, both of animal and vegetable origin.

The animals are omnivorous and their diet eventually includes aquatic insects and their larvae, algae, nuts, grain, seeds, fruits, small molluscs and crustaceans, fish flesh and offal of fish and other animals. They can also be cannibalistic. Further reading on this topic may be found in Ling (1969b).

2.3 Morphology

The following observations on the external anatomy of the freshwater prawn, *M. rosenbergii*, are taken from the work of Ling (1969a), except where noted.

The eggs are slightly elliptical, of long axis 0.6-0.7 mm, bright orange in colour, until 2-3 days before hatching when they become grey-black.

The larvae go through 8 (Ling, 1969a) to 11 (Uno and Soo, 1969) distinct stages before metamorphosis. Each has several distinguishing features. A simplified key to these stages is given in Appendix 5. The first stage larva is just under 2 mm long (from the tip of the rostrum to the tip of the telson) while, by the eleventh stage, it has grown to over 7 mm long (Uno and Soo, 1969).

The newly metamorphosed post-larva is also about 7 mm long and is characterized by having locomotion and swimming behaviour similar to that of an adult. It is generally translucent and has a light orange-pink head portion.

The older juveniles and adults of *M. rosenbergii* are normally distinctively blue, sometimes brownish in colour (they are not red until they are cooked). The second of the five pairs of walking legs is very much larger than the others and ends in a more pronounced claw. Both of these legs are equal in length (unlike some other species of *Macrobrachium*).

Mature male prawns are considerably larger than the females and the second walking leg is much larger and thicker. The cephalothorax ('head') of the male is also proportionately larger, and the abdomen is narrower than the female. The genital pores of the male are between the bases of the fifth walking legs. An alternative technique for sexing juvenile prawns is shown in Figure 1(g).

The head of the mature female and its second walking legs are much smaller than the adult male. The genital pores are at the base of the third walking legs, the pleura of the abdomen is longer and the abdomen itself is broader. The pleura form a broad chamber in which the eggs are carried between laying and hatching. A ripe or 'ovigerous' female can easily be detected because the ovaries can be seen as large orange-coloured masses occupying a large portion of the dorsal and lateral parts of the cephalothorax.

Photographs of the adult male, adult female and of the egg-carrying female are shown in Figure 1, and diagrams of the technical terms used above are given. It is worth mentioning here that many people find it hard to distinguish between *Macrobrachium* and penaeid[1/] (marine) prawns once they have been harvested and the heads have been removed. If the 'tail' still retains its shell there are, in fact, two easy ways of distinguishing them (Fincham and Wickins, 1976). *Macrobrachium* spp. have a smooth rounded dorsal surface to the abdomen while penaeids have a simple or complex ridge at the dorsal apex of the abdomen (Figure 1(e).

1/ Many of the large species of prawns internationally marketed belong to this group, including the US Gulf Coast white, pink and brown shrimp, the Kuruma shrimp of Japan, the sugpo (jumbo tiger shrimp) of the Philippines, the banana prawn from Australia.

Additionally, the second pleuron of the abdomen (or tail) of *Macrobrachium* (in common with all caridean prawns, including some marine shrimp such as *Crangon* spp., *Pandalus* spp., and *Palaemon* spp.) overlaps both the first and the third pleuron. In penaeids the second pleuron overlaps the third pleuron only and is itself overlapped by the first (Figure 1(e).

3. HATCHERY SITE REQUIREMENTS

3.1 Water

The establishment of a freshwater prawn hatchery normally requires a coastal site, although there are some alternatives (Manual Section 5.8.5). Abundant fresh water must also be available. The ideal site, from the point of view of water supply, which is of paramount technical importance, is one where, by sinking wells to different depths, both fresh water and sea water can be obtained from below ground.

Several well-known freshwater prawn hatcheries draw their sea water from wells sunk in the natural underlaying coral of their site, which seems to provide pollution-free supplies. If this type of site is not available, one which has direct access to a sandy beach with mixed sand particle size should be chosen. On this type of site a shallow beach 'well' of the type described in Appendix 1 can be utilized.

Ideally, fresh water should also be obtained from underground sources for hatchery use. City tap water is also normally suitable, provided it is vigorously aerated, say, for 24-48 hours before use, to remove residual chlorine. Well water should also be aerated by cascading for example, to bring its dissolved oxygen level up to saturation or near.

Many freshwater prawn hatcheries utilize surface supplies for both fresh water and sea water, a practice which cannot be recommended. These hatcheries draw sea water from a rigid pier offtake in the sea or a flexible buoyed system. Crude screening is necessary to prevent ingress of the larger flora and fauna. Fresh water is also often supplied by gravity or pumped from surface supplies such as rivers or irrigation ditches. This practice exposes the hatchery to severe variations in water quality and particularly to water contamination from agricultural chemicals. Estuarine water varies in salinity both diurnally and seasonally.

If sea water or fresh water is drawn from surface supplies, some form of filtration is necessary. Normally this involves some form of gravel/sand bed filter. Two forms of filters are described in Appendixes 1 and 2. Because of the extra problems and dangers involved, the authors do not recommend the siting of freshwater prawn hatcheries in areas where only surface water supplies are available. Experience shows, however, that this will not prevent the development of such hatcheries. The minimum requirement during site evaluation should be to carry out watershed surveys and water analyses, especially for pesticides.

Both fresh water and sea water used for hatchery purposes should have a pH in the range 7.0-8.5 pH and a temperature as close as possible to the optimum range (28°-$31^{o}C$). Hydrogen sulphide should be absent. If tap water is used, chlorine must be removed by aeration. Small post-larvae are more susceptible than several species of marine shrimp to nitrite and nitrate (Wickins, 1976), both in terms of acute and chronic toxicity (the latter resulting in poorer growth and survival). Armstrong, Stephenson and Knight (1976) also report sublethal effects of nitrite at levels as low as 1.8 ppm (NO_2-N) with larvae of *M. rosenbergii*. Tentatively we suggest that hatchery intake water should not have levels of nitrite and nitrate higher than 0.1 ppm (NO_2-N) and 20 ppm (NO_3-N).

Sea water should have as little diurnal or seasonal variation as possible. Apart from indications that results are better in freshwater hatcheries using fresh water having total hardness of less than 100 ppm $CaCO_3$, almost nothing is known about the characteristics of the 'ideal' water supply. Typical analyses of fresh water at two successful hatchery sites are given in Table 1. These show very low levels of iron. Manganese should also be low. Clearly a hatchery should not be sited, especially where surface water is used, where its water supplies are endangered by pollution from tanker discharge, oil refineries, tanning, agricultural pesticides and herbicides, or chemical plants, for example.

(a) Adult male M. rosenbergii

(b) Adult male and female M. rosenbergii

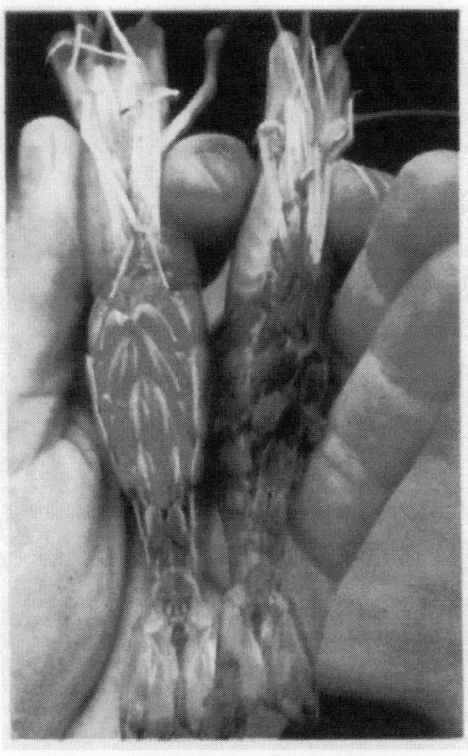

(c) Egg-carrying ("berried") female M. rosenbergii

Figure 1. External appearance of freshwater prawns

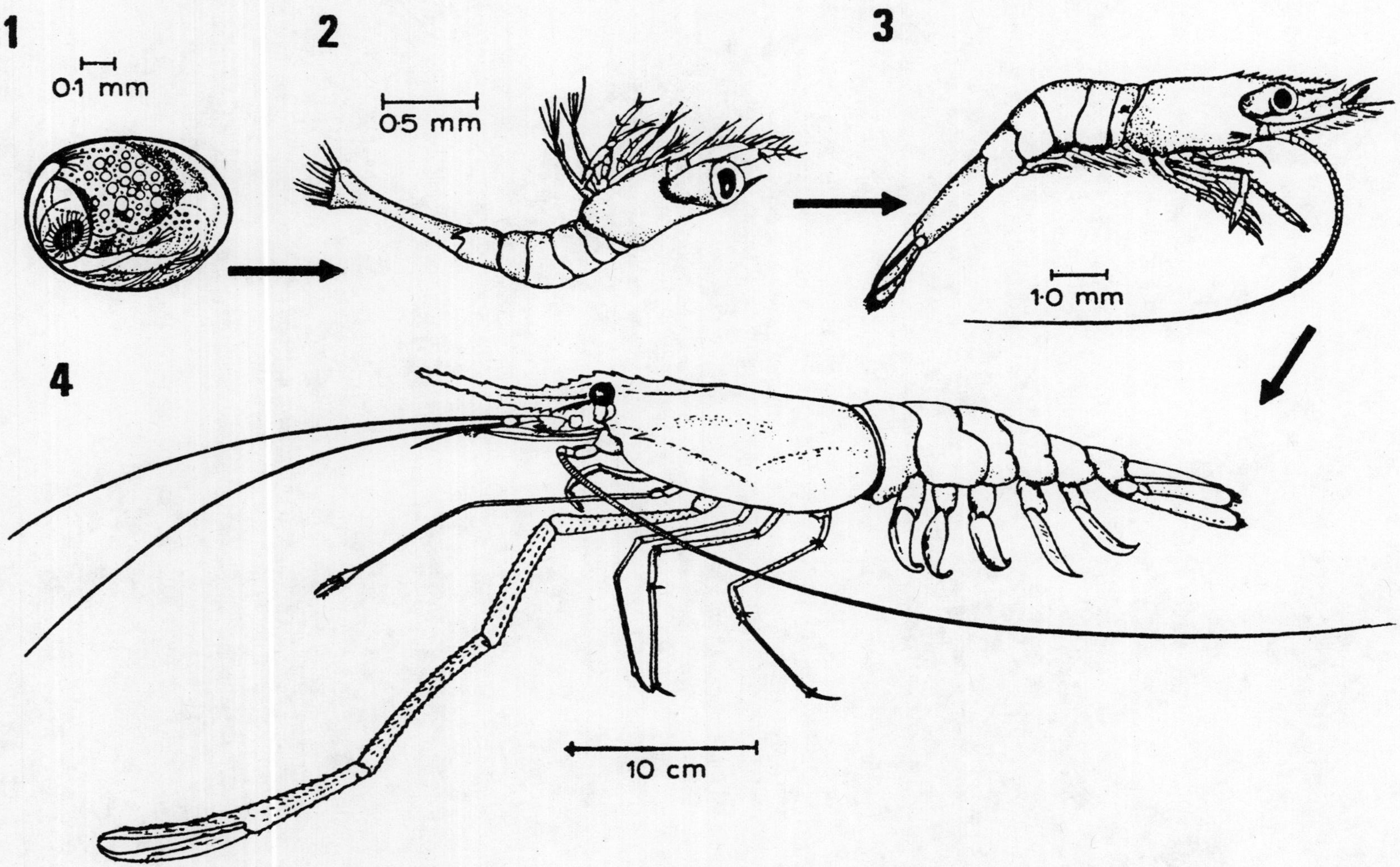

Figure 1 (continued). (d) The life cycle of a caridean prawn: 1. egg; 2. larva; 3. postlarva; 4. adult. (Forster and Wickins, 1972)

(i) Cross sections of abdominal segment 6[2/]

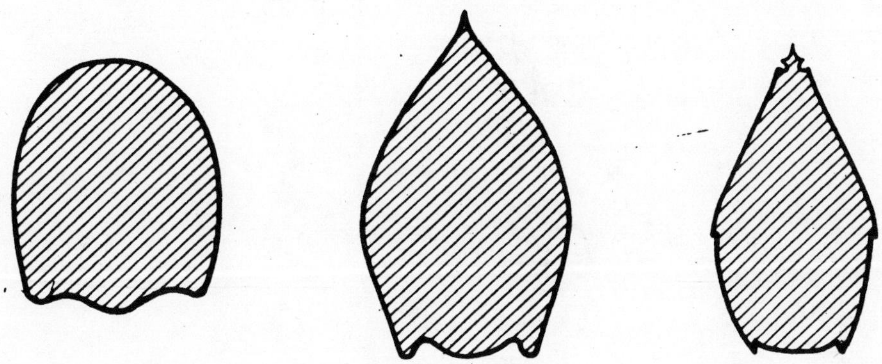

(ii) Side view of abdomen (head removed; pleopods not shown) showing difference in overlapping of pleura

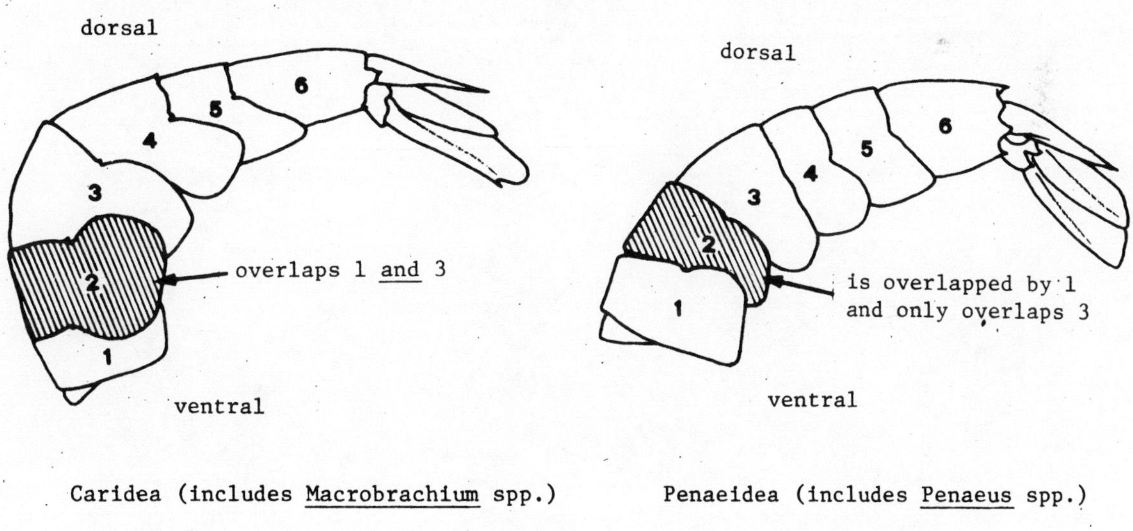

Figure 1 (continued). (e) Distinguishing characteristics between Macrobrachium[1/] spp. and Penaeus spp. (Fincham and Wickins, 1976)

1/ The characters portrayed for Macrobrachium are true for other caridean prawns too, some of which (e.g., Pandalus, Crangon, Palaemon) are marine.

2/ Last segment before telson.

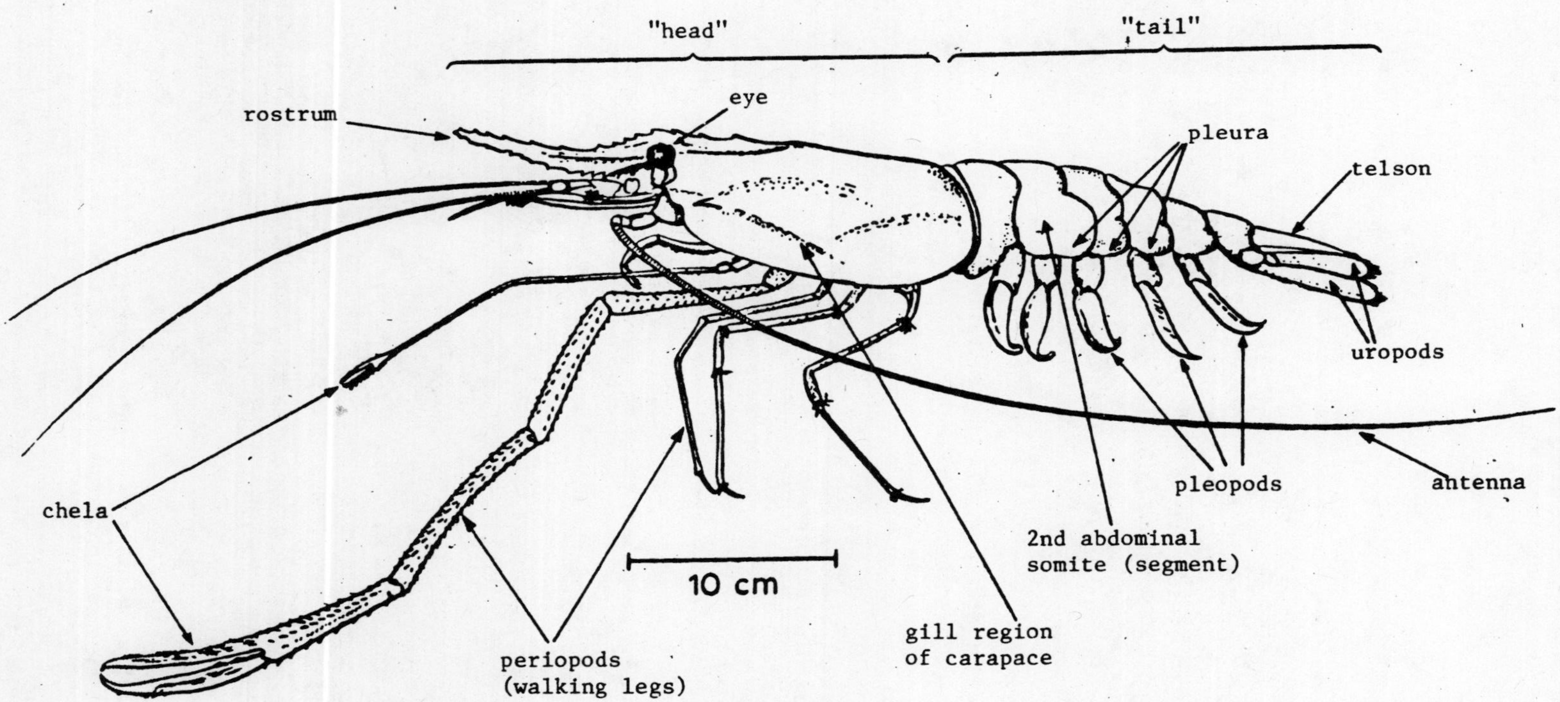

Figure 1 (continued). (f) Some anatomical features of the freshwater prawn. Drawing based on Forster and Wickins (1972)

(i) Medium cross section of first abdominal somite of male Macrobrachium rosenbergii

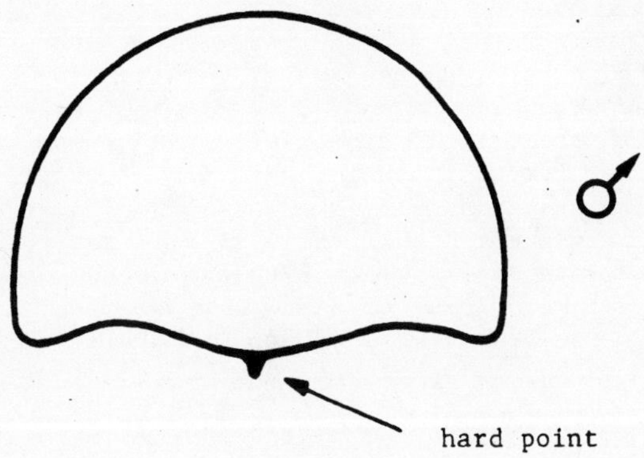

(ii) View of ventral side of first abdominal somite (pleopods not shown)

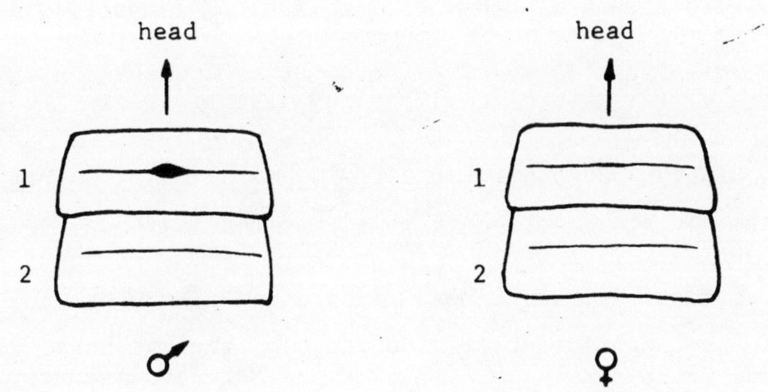

Figure 1 (continued). *(g)* Aid to the sexing of immature freshwater prawns

Examine the ventral side of the first somite (segment) of the abdomen. Males have a lump or point in the centre of the somite, which can be felt with the finger.

Although high iron levels seem detrimental, Ferdinando and Manawadu (1982) have reported successful larval prawn rearing using well water with an original soluble iron content of 15-20 ppm. Atmospheric oxidation and the activity of iron bacteria after 48 hours storage in open tanks was sufficient to precipitate the iron. After pumping through a pressure sand filter the soluble iron content was down to less than 2 ppm. A simple method of cascading to reduce the iron content of water is described in Cansdale (1979).

The quantity of fresh water and sea water required for a freshwater prawn hatchery depends not only on the proposed scale of operation but also on the salinity of the sea water, which controls the proportions of sea water and fresh water necessary to produce $12^{o}/oo$ brackish water for larval rearing (Table 2). The consumption of $12^{o}/oo$ water for each 10-m^3 tank (the size used as an example in this manual) averages 4-6 m^3 (4 000-6 000 litres) per day. Pumping capacity must be sufficient to fill the tank with brackish water within one hour in order to make the daily water exchange as rapid as possible. Thus, though each tank will consume an average of less than 4 litres/min, pumping and pipework capacity must be sufficient to supply the peak demand of approximately 170 litre/min per 10 m^3 tank (in case of emergencies in which all water has to be changed, see Manual section 5.2.4). Put another way[1/], up to 180 m^3 of $12^{o}/oo$ water are consumed for every 100 000 post-larvae produced without problems. In addition, sufficient additional fresh water to maintain holding tanks for post-larvae should be allowed for. For a hatchery operating five 10-m^3 larval tanks, an additional consumption of up to 42 m^3/day of fresh water should be provided for the supply to post-larval holding tanks. On average the consumption of brackish water by a hatchery of this size (five 10-m^3 larval tanks) would be 20-30 m^3/day.

Techniques which decrease the water consumption or freshwater prawn hatcheries are dealt with in Manual Section 5.8.

3.2 Other Requirements

A good hatchery site should also have the following characteristics:

(a) a secure power supply which is not subject to lengthy power outages. An on-site emergency generator is essential for any hatchery where the scale of investment warrants it;

(b) good all-weather road access for incoming materials and outgoing post-larvae;

(c) not more than 16 hours transport time from the furthest farm it is to supply by land transport;

(d) access to professional biological assistance from government or other sources;

(e) land, of an area appropriate to the scale of the hatchery, with good access to seawater and freshwater supplies. The cost of pumping water supplies to a site elevated high above sea level is an important factor in the economics of the project;

(f) a climate which will maintain water in the optimum range of 28°-30°C without costly environmental manipulation;

(g) access to food supplies for larvae (Appendixes 3 and 4);

(h) a high level of technical and managerial skills.

Further reading on the topic of site selection generally and particularly applied to freshwater prawn culture is given by Webber (1973), New (1975) and New *et al.*(1977).

4. HATCHERY FACILITIES

Hatchery design will depend on the scale of production desired, the characteristics of the site (such as topography, climate, etc.), the type of building materials obtainable locally and the finance available. It is not therefore intended in this manual to provide

---

1/ Assumes a production of 100 000 post-larvae per 10-m^3 tank in a 30-day period. Actual production should be better, thus reducing water consumption per post-larva produced

a complete hatchery design for freshwater prawns but merely to describe certain features. Only one type of hatchery is described here: one which has proved effective in Thailand and which was based upon the original facilities at the Anuenue Fisheries Research Centre in Hawaii. There are many other successful techniques and facilities, some of which are mentioned in Manual Section 5.8. The system which we describe is simple to construct and operate. No special facilities for holding broodstock are described as berried females are placed directly into the larval tank (Manual Section 5.1).

4.1 The Larval Tank

The hatchery centres around the larval rearing tank. Many different types of containers are used to grow freshwater prawn larvae, including circular flatbottomed tanks (made from plastic or converted from large-bore drain pipes), circular conical-bottomed plastic tanks, plastic-lined wooden tanks and even 'klong pots' (earthenware water jars). We have found that rectangular tanks are the most practical to use. Small circular tanks are good to use too, but once you scale up the operation, you either have to have very large diameter tanks or a large number of small tanks. Both become cumbersome to operate and the latter is expensive in terms of space, the number of pipe fittings required, etc. The principal advantage of a rectangular tank is that only its length alters as its size is increased. A 10-m^3 tank is just as accessible for feeding, cleaning and larval inspection as a 1-m^3 tank.

The materials most suitable for tank construction will vary from site to site. In chosing materials for tank construction and for pipes, pumps, etc., it should be noted that copper and zinc (and their alloys), galvanized steel, bare concrete and oil have been noted as toxic to larval freshwater prawns. If cheap enough, rigid plastic, fibreglass or plastic-lined wooden tanks will do. The original 'Hawaiian' tanks were fibreglass inside, with a reinforced layer of concrete 'shot-creted' onto the outside for strength. Where they are cheapest, as in Thailand, good quality concrete or concrete-faced hollow-block constructed tanks are ideal. Hollow-block tanks should be reinforced with vertical iron rods. In both cases the interior surface should be coated with several layers of pure epoxy resin to prevent harmful chemicals leaking out of the concrete and to provide a smooth surface. Tanks with dark coloured interiors seem to give better results, though not all successful hatchery operators would agree with this statement. This smooth surface and the 'rounding off' of all right-angled parts of the tank (where the side walls and the bottom meet) facilitates tank cleaning, reduces the surface available for algal, bacterial and protozoal growth and strengthens the tank where reinforced concrete is not used. Whatever materials are used to construct tanks, it is essential to 'age' new tanks by soaking them in several changes of brackish water for several weeks. This allows soluble toxic materials to leak out. This applies also to epoxy resin coated tanks.

A convenient internal size for these rectangular tanks is 14 m^2 which, with a water level of 70 cm holds approximately 10 m^3 of water. Internally the tanks are 1 m deep, 2 m wide and 7 m long. Water and air intake supplies are positioned at one end of the tank and the drain at the other (Figure 2). The tank bottom is sloped slightly toward the drain end. A 4-in (10 cm) turn-down drain is utilized, which is protected by a filter sock (Figure 3) to prevent the loss of larvae during water-changing operations. Vigorous aeration is supplied to the tank through two or three rigid PVC pipes or weighted flexible plastic tubing laid on the bottom of the tank parallel to its long axis (1/2-in (1.2 cm) pipe is adequate for a 10-m^3 tank). Pipe or tubing with small holes punctured in it seems to give fewer problems than airstones.

Larval tanks should be carefully constructed so that they do not leak. This means they must have a firm, well-compacted foundation since a 10-m^3 tank will carry 10 t of water (plus its own weight). Concrete pouring or facing work must be continuous so that the concrete does not dry out by sections. Failure to do this will result in cracks and leaks later, at the joints.

Larval tanks should be positioned sufficiently high so that they will drain by gravity when the turn-down drain is operated. Tile, faced block or concrete drainage canals must be constructed to carry away the drained larval-rearing water without undermining the foundations of the tanks.

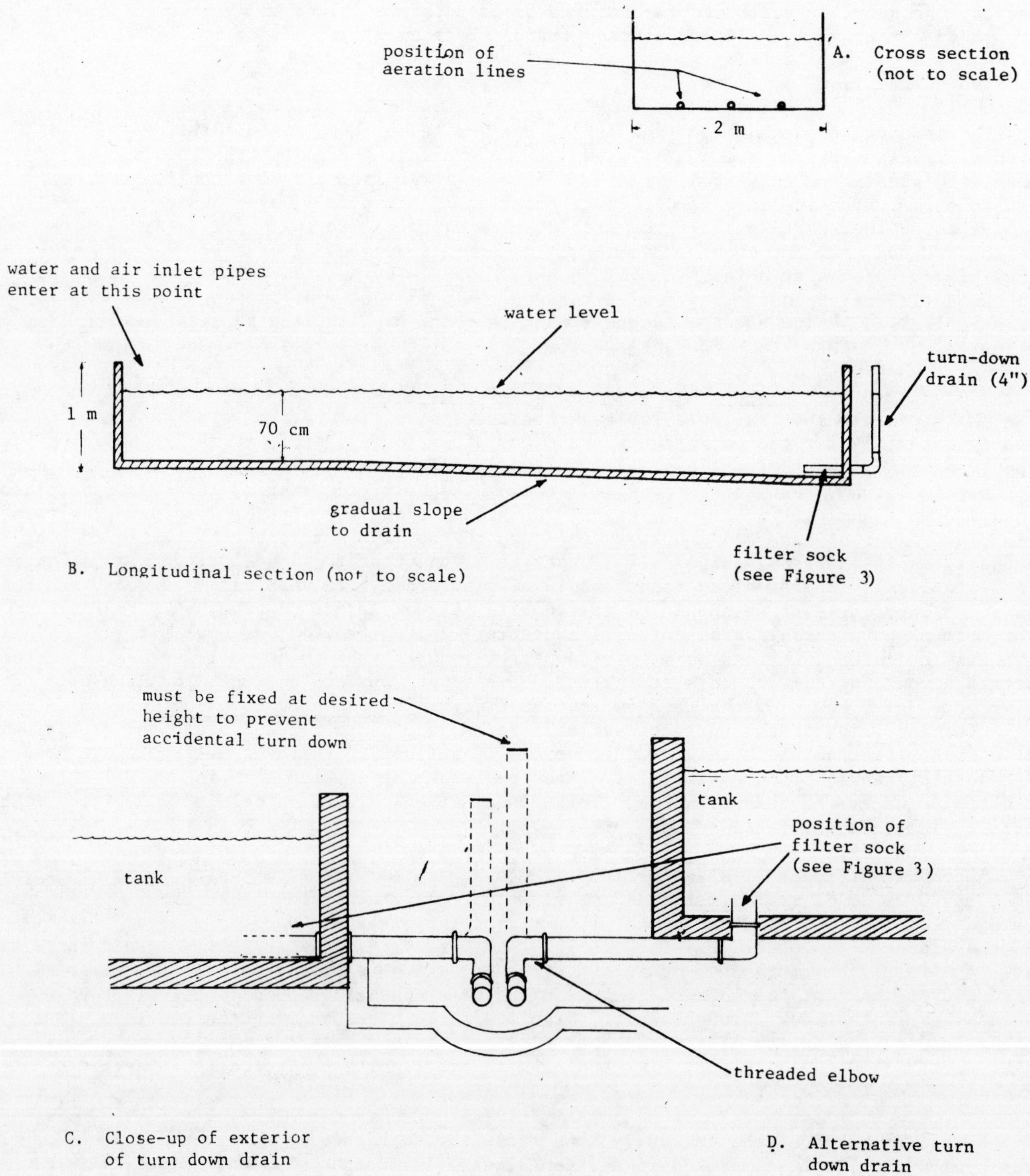

Figure 2. Larval tank design

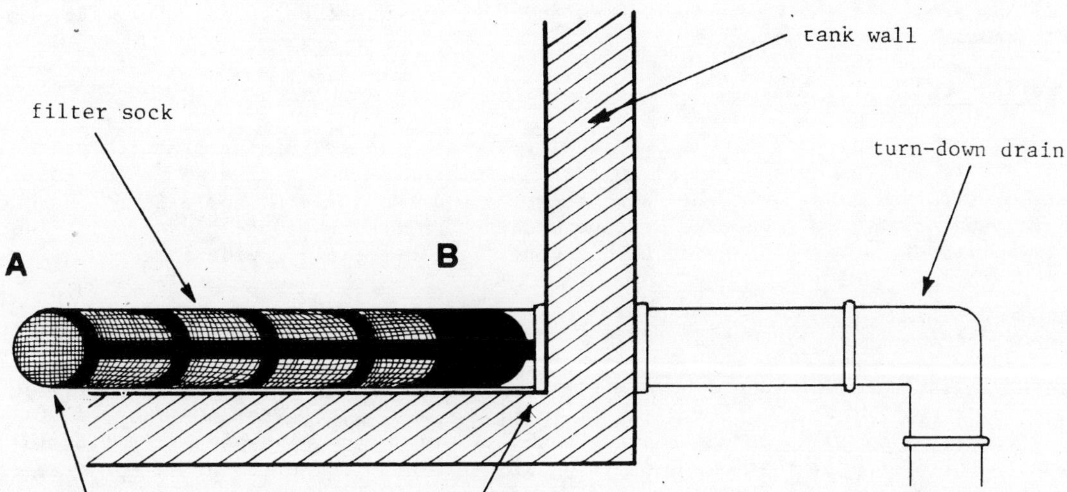

Filter sock consists of a 60 cm long piece of 4" PVC pipe which slots into a push fit joint close to the tank wall or flush with the tank bottom. The filter is more easily removed from its seating if it has a number of small holes bored in it at the end "A", through which a hook can be inserted. The PVC pipe has most of its surface area cut away and the whole thing is covered with mosquito netting which is securely stuck down with PVC glue at points "A" and "B" and along the axis between these two points and tied at points "A" and "B". The filter sock can be removed for cleaning or replacement with one with a larger mesh. Initial mesh hole size should be not greater than 250 µm and can gradually be increased to 1 200 µm for postlarvae.

Figure 3. Filter sock for larval tank

Tanks can be in the open but simple shading (palm fronds or a bamboo framework, for example) should be provided where there is a possibility of water temperature rising too high. Conversely, the tanks can be housed in a greenhouse-type structure to bring the water temperature up to the optimum level where air temperatures are low during all or part of the year. If the tanks are housed indoors, care must be taken to provide some sunlight (Manual Section 5.2.5).

4.2 Holding and Mixing Tanks

Tanks for holding post-larvae before distribution and for mixing brackish water are also necessary for the hatchery. As in the case of larval tanks, the method of construction, size and shape will vary with the site and the scale of operations. We have found that units of 50-m^3 concrete or concrete-faced block tanks are convenient. Their design is similar to a larval-rearing tank except that they are 7 m wide instead of 2 m, and 10 m long instead of 7 m.

If possible, those tanks used for mixing brackish water, or for storing sea or fresh water, should be elevated so that 12 ppt water mix can be distributed to the larval tanks by gravity. The hatchery should have a total storage, holding and mixing capacity of about four times the volume of its larval rearing tanks (e.g., four 50-m^3 or two 100-m^3 tanks for every five 10-m^3 larval-rearing tank). This capacity is necessary to allow for adequate water storage, treatment and mixing time for the production of 12 ppt brackish water (Manual Section 4.4) and to provide space for storing post-larvae before distribution (Manual Section 5.5).

4.3 Air

An oil free blower is better than an air compressor to supply air to the hatchery, for it gives high volume, low pressure uncontaminated air which is what is required. The high pressure provided by an air compressor is not normally required, except for flushing filters, if these are used (Appendix 2). A 200 CFM (5.6 m^3/m Roots-type or similar blower) is sufficient to supply air for a hatchery capable of producing 20 million larvae/year.

Every larval, holding and water-mixing tank must have an adequate air supply, as must brine shrimp hatching containers. It is important to ensure that the flow of air in one tank is not affected by the number of other tanks in operation or by the operation of valves on an adjacent tank. This can be achieved by having a large bore 2 or 3 in (5 cm or 7.5 cm) ring main distribution system (Figure 4) with smaller 1/2 in (1.2 cm) or one in (2.5 cm) in standpipes for each tank. The blower should be oversize for the maximum usage and excess air should be voided through a valve on the ring main which can be adjusted according to the day-to-day requirements of the hatchery.

The aeration system is a vital part of the hatchery. The distribution system should be buried and protected from accidental damage. This can be achieved by placing it under 4 in (10 cm) of medium gravel or sand, whether the hatchery is indoors or out-of-doors. Covering any type of hatchery pipework with concrete is not recommended.

<u>A spare blower and motor, in working order, should be available at all times.</u> Monthly rotation of the blowers is good practice, as are daily checks that both blowers are in working order. Preferably the spare blower should be permanently plumbed and wired in so that it can be switched into the system immediately if a failure of the other blower should occur. A sophisticated hatchery may even have a pressure drop sensor built into the air distribution system which will switch the emergency blower on automatically should failure of the other one occur.

4.4 Water

4.4.1 Supply

The desirable characteristics of sea water and fresh water supplies have been dealt with in Manual Section 3.1.

4.4.2 Treatment

The following water treatment schedule is suggested for surface or shallow well water supplies:

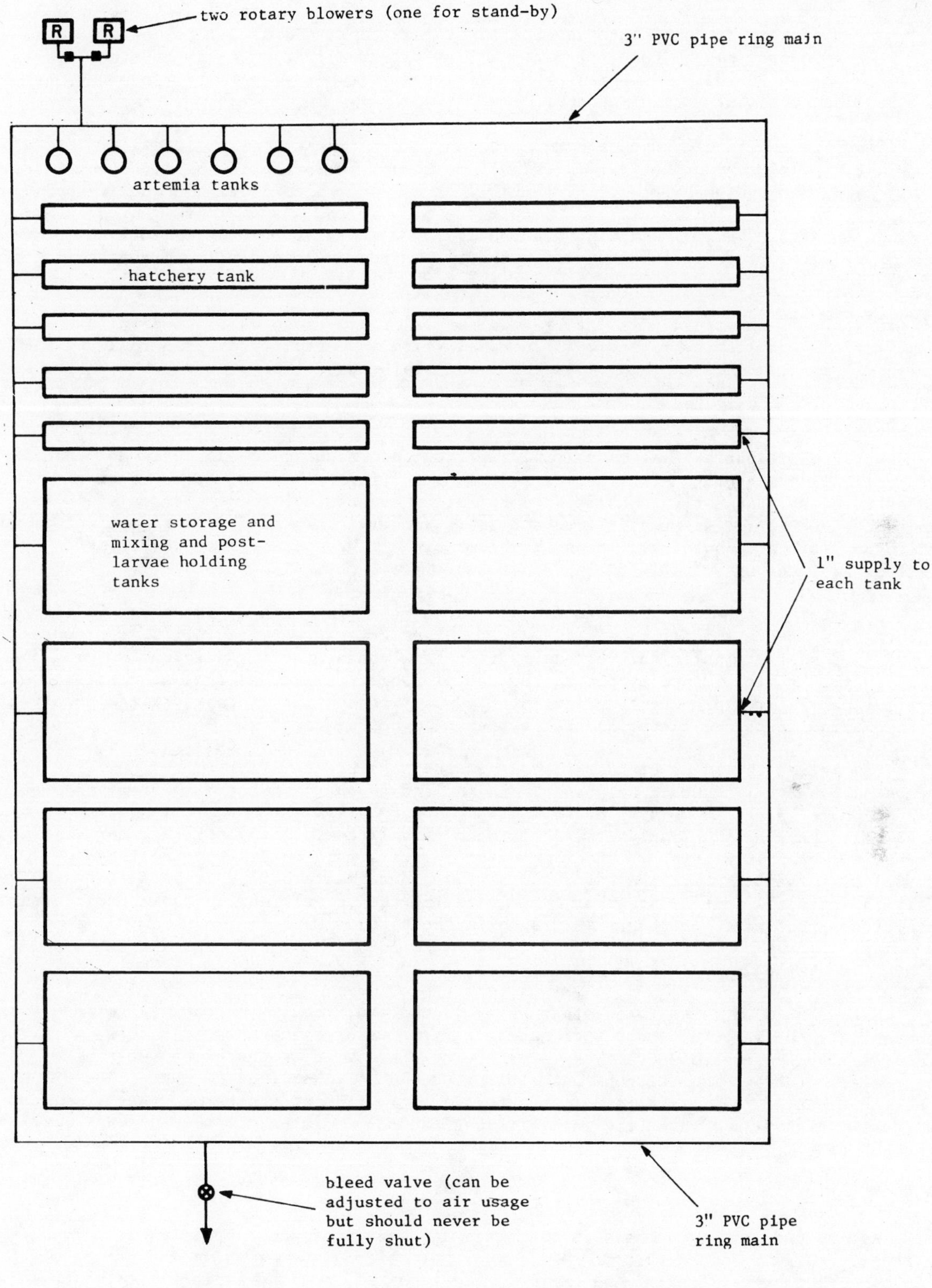

Figure 4. Hatchery air distribution system (tanks not to scale)

(i) Sea water:[1] pump into an aerated storage tank; add 25 ppm formalin (see Appendix 9 under 'ppm' for method of calculation); allow to precipitate over a 6-day period; allow precipitate to settle without aeration for one day; pump supernatant to mixing tank.

(ii) Fresh water:[1] pump into an aerated storage tank; add 6 ppm commercial bleach (calcium hypochlorite, $Ca(OCl)_2$), or 60 ppm 'chlorox' solution (approximately 5.25% sodium hypochlorite (NaClO)), either of which provide about 1.5 ppm chlorine [2]; allow to precipitate for 5 days; add 10 ppm sodium thiosulphate ($Na_2S_2O_3.5H_2O$), aerating vigorously for one day to remove residual chlorine; allow precipitate to settle without aeration for one day; pump supernatant to mixing tank.

(iii) Brackish water: mix the treated sea water and fresh water to a salinity of 12°/oo (Table 2), store (if necessary) in a constantly aerated tank and pump to larval tanks as required after the salinity and temperature have been checked.

4.4.3 Distribution

Water distribution systems vary widely from one hatchery to another. Many hatcheries are built with elaborate distribution systems supplying sea water, fresh water and brackish water to every tank, only to allow them to fall into disuse. These systems are often abandoned because of in-line pump failure and fears about water quality due to the use of water which has lain stagnant in the pipes for some time. The in-line water distribution system is then replaced by flexible tubing and submersible pumps.

Submersible pumps are easy to use if the hatchery is compact, but their indiscriminate use results in contamination between untreated and treated water supplies, and, worse still, possible disease transfer between one rearing tank and another. An inline pumping and distribution system is preferable to the use of submersible pumps and, as mentioned earlier, gravity can replace pumping for distributing water to the larval tanks if the brackish-water mixing tanks are elevated. The system need not be complex.

Pump sizes should be standardized as far as possible to minimize the number of standby pumps necessary. It should be possible to replace an out-of-order pump simply and quickly, and adequate spares in working order must be on site at all times. It is not necessary to have more than one supply pipe/valve for each larval tank. A conceptual water distribution layout is shown in Figure 5. The tank water inlets should be designed so that they can be turned away from the tank and water can be flushed to waste before directed into the tank. This obviates the fear of stagnant water entering the larval-rearing tanks and prevents the use of very hot water which has warmed up in the pipes.

4.4.4 Discharge

Care should be taken to see that water discharged from the hatchery after use does not contaminate the incoming sources of hatchery fresh water and sea water. This is particularly important where surface sources of water are utilized. In a coastal hatchery using surface sea water, tidal and current characteristics should be taken into account in determining the locations of the intake in relation to the farm effluent discharge. Where surface fresh water is taken from a river the farm effluent should be discharged well below the water intake point.

4.5 Pumps

Again, pump sizing depends on the scale and design of each specific hatchery. Specific hatchery design is not a part of this manual. In sizing, pumps should be chosen which will fill the appropriate tank at the maximum rate required, not the average rate. There is nothing more annoying than a slow filling tank due to pump undersizing.

1/ If the water is unfiltered and turbid it may be necessary to allow solids to settle out in one tank before pumping the supernatant to another tank for treatment

2/ The use of bleach instead of 'chlorox' (at least in Thailand) has proved cheaper but commercial bleach has many impurities and has been suspected of causing water toxicity

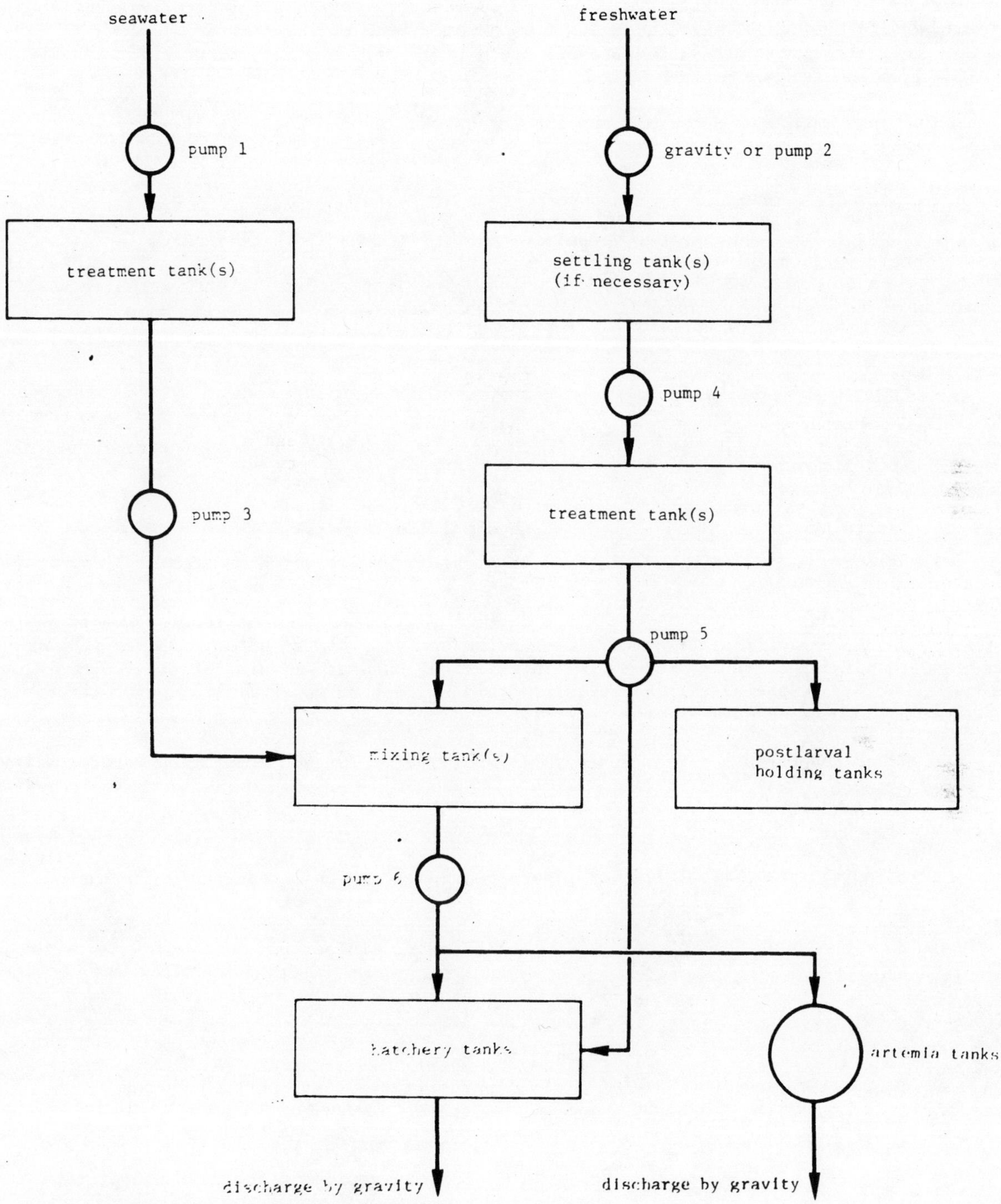

Figure 5. Example of hatchery water distribution layout

Notes: 1. Pumps 1 and 2 must be permanent

2. Pumps 3, 5 and 6 can be the same submersible pump

3. Pump 4 can also be a submersible pump

4. Submersible pumps used for untreated water should not be used for treated water and vice versa

5. Pumps 5 and 6 can be eliminated if the mixing tanks are constructed to allow for gravity feed to the larval tanks

Copper and zinc are toxic to freshwater prawns but there should be no problem in the use of pumps containing alloys of these two metals (which are often chosen, particularly for seawater pumping, because of their corrosion resistance) where the water passes through the pump only once. Pumps which are submerged in water, or which are part of recirculation systems, should have those parts in contact with water made of inert material, such as plastic. Air lift pumps are also extremely useful for recirculating water or for transfer of water from one tank to another.

Useful information on pumps and pumping for hatcheries and ponds is given by Jamandre (1977).

4.6 Monitoring Water Quality

A freshwater prawn farm should be able to monitor the water quality parameters listed below. Specific equipment for these analyses is not identified in this manual, since availability varies so much from location to location. The list applies to both hatcheries and production sites with the exception of item (d) which is for pond use only (Manual Section 8.3). The facility to make the following determinations is a basic requirement.

	Parameter	Technique
(a)	Temperature	Thermometer } or electrical (portable meters)
(b)	Dissolved oxygen	Chemical } or electrical (portable meters)
(c)	pH	Chemical } or electrical (portable meters)
(d)	Turbidity	Physical (Secchi disk)[1/]
(e)	Salinity	Physical (Portable refractometer or hydrometer)

It is not practical for small farms to install facilities for the other types of analytical work, especially where they involve the analysis of sea water or brackish water. Samples of water for the analysis of other parameters, such as ammonia, nitrite, nitrate, hardness, metals, pesticide residues, etc., should be sent to governments, universities or private laboratories who have the facilities and staff to deal with them.

Strickland and Parsons (1968) and Boyd (1979) provide further reading on water quality and analysis.

4.7 Miscellaneous

The following is a partial list of general equipment which is necessary for the hatchery:

(a) Buckets
(b) Scales
(c) Nets
(d) Mesh
(e) Flexible tubing
(f) Spares for electrical equipment
(g) Spares for PVC pipework and valves
(h) Epoxy-resin paint
(i) Fibreglass repair kits
(j) Tools
(k) Brushes
(l) Post-larval transport equipment (bags, tanks, oxygen bottles, etc.)
(m) Disease prevention drugs and chemicals
(n) Kitchen equipment for feed preparation

5. HATCHERY OPERATION

5.1 Egg Supply and Hatching

Freshwater prawn eggs are carried on the lower side of the abdomen of the adult female prawn and are easily visible (Figure 1(c). Adult animals at this stage are known as "berried females".

1/ See Figure 6

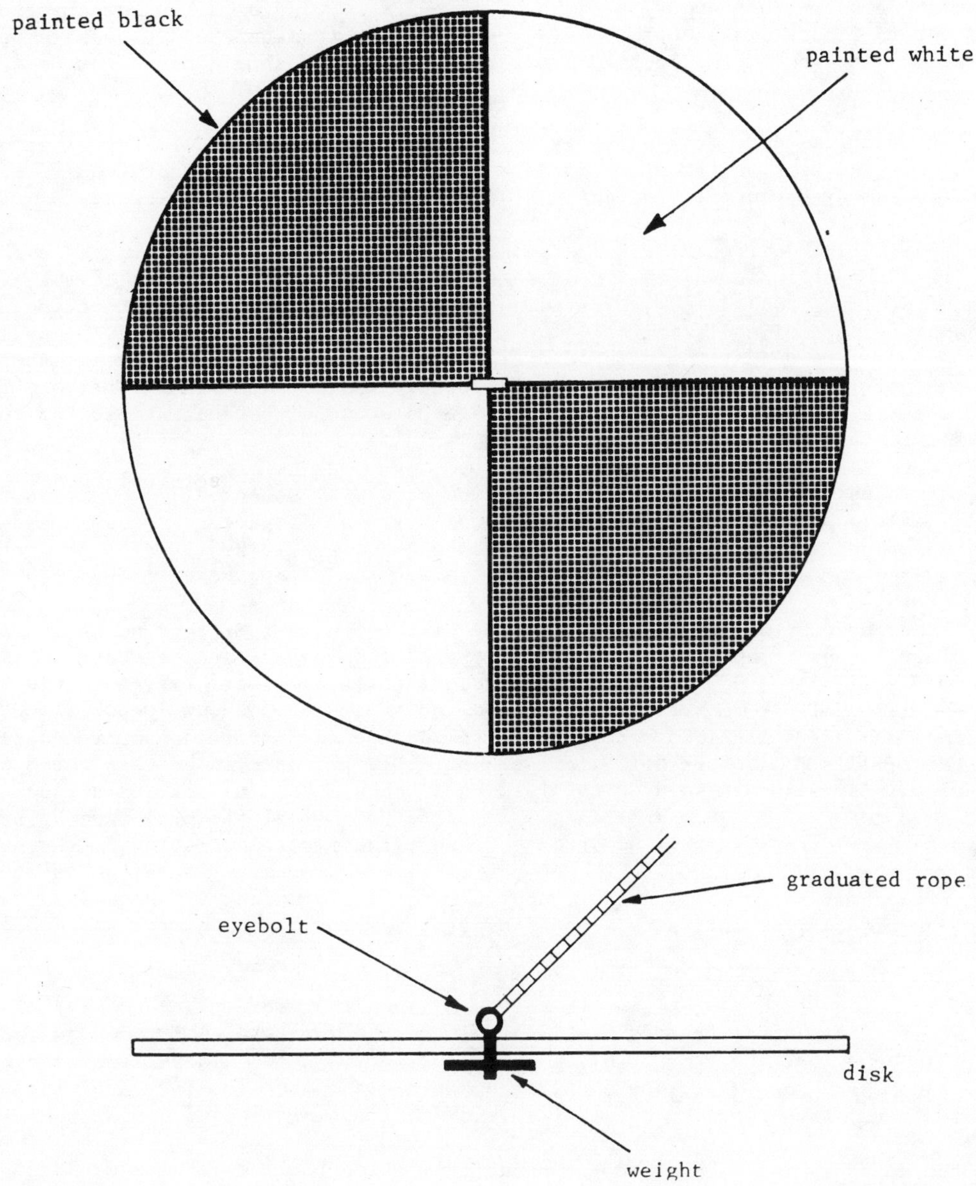

Figure 6. Secchi disk

Berried females for hatchery purposes can be obtained from rivers, farm ponds, or from breeding stock maintained and mated in aquaria. If *M. rosenbergii* is not indigenous in the country in which it is to be farmed, berried prawns can be imported using techniques similar to those used for transporting post-larvae (Manual Section 5.6), but blunting the rostrum with scissors to prevent the bags being punctured. From a hygiene point of view it is better to import disease-free post-larvae, rather than berried females. The permission and assistance of the local Department of Fisheries should be sought on this topic.

Normal practice, once freshwater prawn farming is established, is for a hatchery to obtain its berried females from farm ponds. Berried females can be obtained from ponds by cast netting but are frequently selected at times of partial or total harvest.

Under natural conditions there are seasonal peaks in breeding activity, normally associated with the beginning of the rainy season, but berried females are always available in a pond containing adult stock. Providing the hatchery has its own production ponds, or maintains a pond specifically for breeding stock, or has a close relationship with the owners of production ponds, there is no difficulty in obtaining berried females whenever required. Their individual value is low, especially as they are usually marketed for human food after the eggs have hatched, so there is no need to economize in the number of berried females utilized for the hatchery.

For hatchery use, berried females should be carefully selected. Animals which are obviously healthy and active, well pigmented and carrying large egg masses should be chosen. Larger females usually carry more eggs; the number required to supply eggs to stock a larval tank depends on the volume of the tank and on the number of eggs carried by each female.

It is not however necessary to exactly predict the number of eggs being introduced into the larval tank if an excess number of berried females is used. A rough guide often used is to assume that 1 000 larvae are produced from each 1 g of berried female weight. Berried females of 10-12 cm (rostrum to telson) normally carry about 10 000-30 000 eggs each. Since many eggs are lost through physical damage and adult consumption during transport of the females, and other eggs fail to hatch, it is recommended that three berried females of this size should be used to stock each cubic meter (m^3) of larval tank water volume. This results, after egg hatching, in an actual larval stocking rate of around 30-50 larvae per litre. As will be stressed later, the important factor is the number of post-larvae produced per tank, not the number of larvae stocked. Where, as is normally the case, berried females are in plentiful supply, it does not matter if the larval tank is overstocked with larvae initially. No attempt at estimating the number stocked is therefore necessary.

When the size of the larval tank being used means that more than one berried female is needed, it is essential to obtain all animals in the same stage of ripeness. Select animals with eggs that are grey or black in colour, not orange. These eggs will hatch within 2-3 days and this will ensure that the larval tank contains larvae of the same age (within 1-3 days) thus reducing cannibalism and facilitating feeding operations.

If the hatchery is adjacent to the ponds from which berried females are being obtained, they can be transported in buckets of water. Berried females travelling greater distances should be transported in the same way as juvenile prawns (Manual Section 5.6). Small bags containing only one animal and transported in darkness will reduce egg losses. Great care should be taken in catching, handling and transporting berried females to minimize egg loss and damage.

On arrival at the hatchery, the berried females should be disinfected by placing them in aerated fresh water containing 0.2-0.5 ppm of copper or 15-20 ppm of formalin for 30 min as a quarantine procedure. Then transfer the females to the larval tank. The females can be fed to demand on post-larval feeds (Appendix 7) but care should be taken not to overfeed, thus causing initially poor water quality for the newly hatched larvae. Often berried females are not fed at all during the 2-3 day period prior to egg hatching.

Egg hatchability is better in brackish water than in fresh water. Some hatcheries allow the eggs to hatch in fresh water for simplicity, raising the salinity after hatching; others place the females in brackish water of 5 °/oo. Berried females will stand the shock of immediate transfer to higher salinities and can be placed, after disinfection, directly into larval tanks with water at the larval rearing salinity (12 °/oo).

Egg hatching, which occurs predominantly at night, can be observed by the presence of larvae in the tank and the absence of eggs on the underside of the females' abdomens. Larvae at all stages can be observed better in the tank with the aid of a white board (Figure 7(b). After the eggs are hatched, remove the spent females from the tank with a coarse mesh dip net.

5.2 Larval Environment

5.2.1 Salinity

Though some hatcheries lower the salinity as the larvae get older, little if any advantage is demonstrated. Larval-rearing salinity is not as critical as many early hatchery operators thought. We recommend maintaining a rearing salinity of 12 °/oo until metamorphosis [1/]. Absolute accuracy is not essential and a range of ± 2 °/oo is adequate. Although slight variation in salinity is not detrimental, sudden wide variations must be avoided. These can occur during water changing when, through operator error, full strength sea water or fresh water is used instead of brackish water. The simplest way to check salinity is by means of a hand-held refractometer.

5.2.2 Temperature

Within a selected temperature range, larvae grow and moult more quickly as temperature increases. The optimum temperature range is 26°-31°C. Below 24°-26°C the larvae will not grow well and the time to attain metamorphosis will be longer. This affects hatchery economics enormously. Similarly temperatures over 33°C are normally lethal, though individual experience in hatcheries tends to vary on this point. Gradual variation in temperature within the optimal range, such as occurs naturally between night and day or cloud and sunshine, for example, is acceptable, though it should be minimized as far as possible. Sudden changes in water temperature must be avoided as they can cause shock and mortality. Sudden changes in temperature, even as little as 1°C, seem to cause trouble and it is therefore essential to have an adequate stock of prepared 12 °/oo water for exchange purposes, maintained under the same environmental conditions as the larval tanks, available at all times. Do not do what one hatchery we observed did, which was to mix incoming fresh water and sea water in the larval tank itself, in this case using sea water from a metal tank exposed to direct sunlight.

If the water level in the larval tanks is too shallow (which happens where operators try to conserve water usage), excessively high temperatures can easily be reached if the tanks are not shaded.

5.2.3 Dissolved oxygen

Oxygen in larval-rearing water should be maintained as close as possible to saturation (Table 3). The aeration system must only be turned off for short periods (for observation of the larvae, for example). It is essential to double-check that air is turned on again immediately after any tank operation which requires its temporary cessation. One of the major causes of larval mortality is operator error on this point.

In practice, if the procedures for water changing, tank cleaning and feeding laid down in this manual are adhered to and there is no failure in the hatchery air distribution system, no problems should be experienced with low oxygen levels. It is not essential to measure dissolved oxygen levels in the larval rearing water, though it would be preferable to do so if a portable DO_2 meter is available. A warning of low oxygen levels would enable the operator to change the water before stress conditions occur (Manual Section 5.2.4).

1/ This salinity applies specifically to *M. rosenbergii*.

(a)

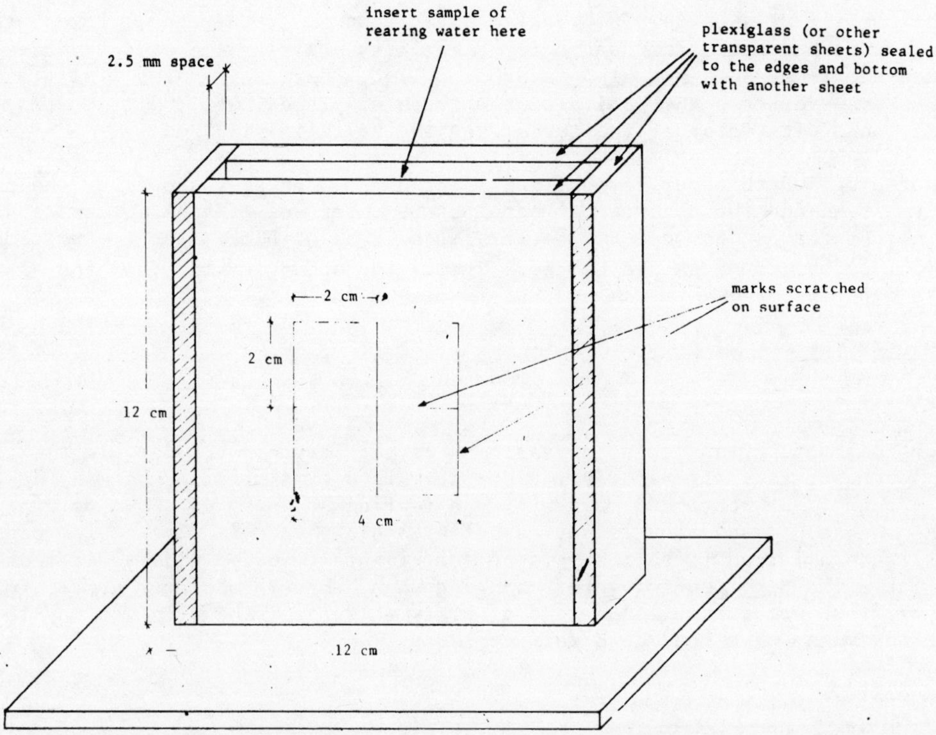

Each of the four squares represents a volume of 1 ml.

By holding the apparatus up to the light, the BSN in each square may be counted and averaged to form an estimate of the density of BSN/ml.

(b)

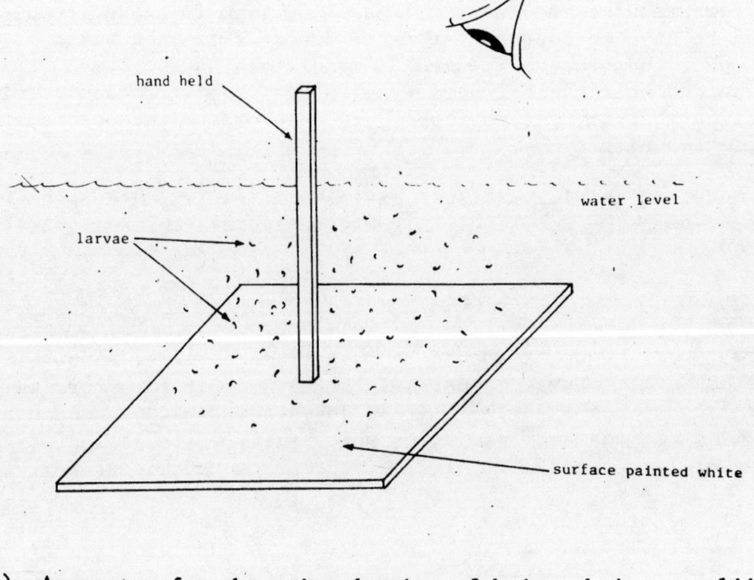

Figure 7. —(a) Apparatus for observing density of brine shrimp nauplii

(b) White board for viewing larvae

5.2.4 General water quality

Many changes in the chemical water quality of larval-rearing water occur which are not visible. These are due mainly to the metabolic wastes produced by the larvae themselves (and brine shrimp nauplii) and by the degradation of excess food. Some of these changes can be extremely harmful to larvae. The most serious are increases in the non-ionized form of ammonia, which is especially evident at high pH, and in nitrite (Manual Section 3.1). It is beyond the scope of this manual to deal with water chemistry but those who wish to study this matter should consult Spotte (1970), Boyd (1979) (this book deals with freshwater chemistry and is more suitable for pond water quality), Wickins (1976) and Armstrong *et al.* (1976)

There are methods which minimize water utilization, such as recirculation/filtration (Manual Section 5.8.4), and maintain good larval water quality. Some proponents of the "greenwater" technique of rearing freshwater prawn larvae claim a beneficial effect on water quality. However, these are sophisticated and not always appropriate techniques. For the simple hatchery, there is no substitute for frequent water exchange, which is what makes the selection of a site with adequate supplies of good quality sea water and fresh water so essential to hatchery success. The hatchery operation described in this manual is based on the "clearwater" technique though reference to the alternative "greenwater" technique is made in Manual Section 5.8.2.

The following procedures are recommended to maintain good larval water quality [1/]:

(a) do not overfeed (Manual Section 5.3);

(b) clean the sides of the tanks every two days by means of a 'squeegie' or scraper;

(c) turn off the air supply to allow solid particles to settle, and siphon off (Figure 8a) surplus food particles and metabolic wastes from the bottom of the tank. This task should be done daily immediately before one of the feeding operations. The time taken to complete this task should be kept to a minimum so that the air can be turned on again as soon as possible. It should be done as part of the daily water exchange procedure. Siphoning will also remove any mortalities which have occurred and provides a good opportunity for the operator to observe the condition of his larvae. Live larvae which are on the bottom of the tank tend to pass through the siphon tube. Some hatchery operators return these to the tanks after collection (Figure 8b); others, believing that the larvae which are on the bottom of the tank and cannot evade the oncoming siphon tube are weak and of poor quality, do not prevent these losses. There is no great danger of losing viable animals until the time of metamorphosis because the larvae swim in the body of the water and do not crawl; thus the latter technique of discarding siphoned larvae is favoured by the authors of this manual;

(d) exchange 50 percent of the water volume every day. This operation should commence 3-4 days after hatching and continue throughout the larval cycle. The quantity exchanged may even be increased to over 50 percent per day toward the end of the rearing cycle, when biomass and feeding levels are at their greatest. The water level should be decreased from 70 cm to about 35 cm through the siphoning operation (see (c) above) and the use of the turn-down drain. The water removed should be replaced by ready-mixed, aerated, 12‰ water at the same temperature as that of the larval tank. This operation should be done before a feeding operation, so that food will not be wasted;

(e) never hesitate to exchange the larval water (in addition to the routine 50 percent per day) at any time when poor water quality is suspected. If the water quality is visibly poor (due, for example, to excessive overfeeding), or smells foul,

1/ See also Manual Section 5.2.6

(a) Siphoning

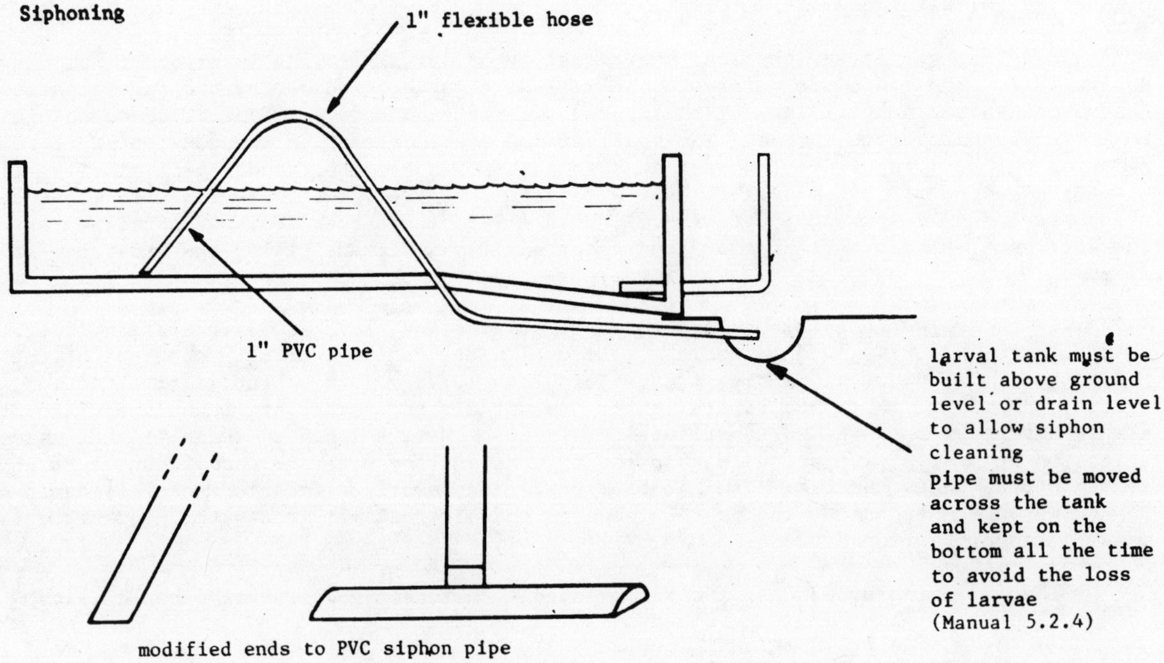

(b) Two forms of collection apparatus to retain any larvae passing through the siphon (exact size is not important)

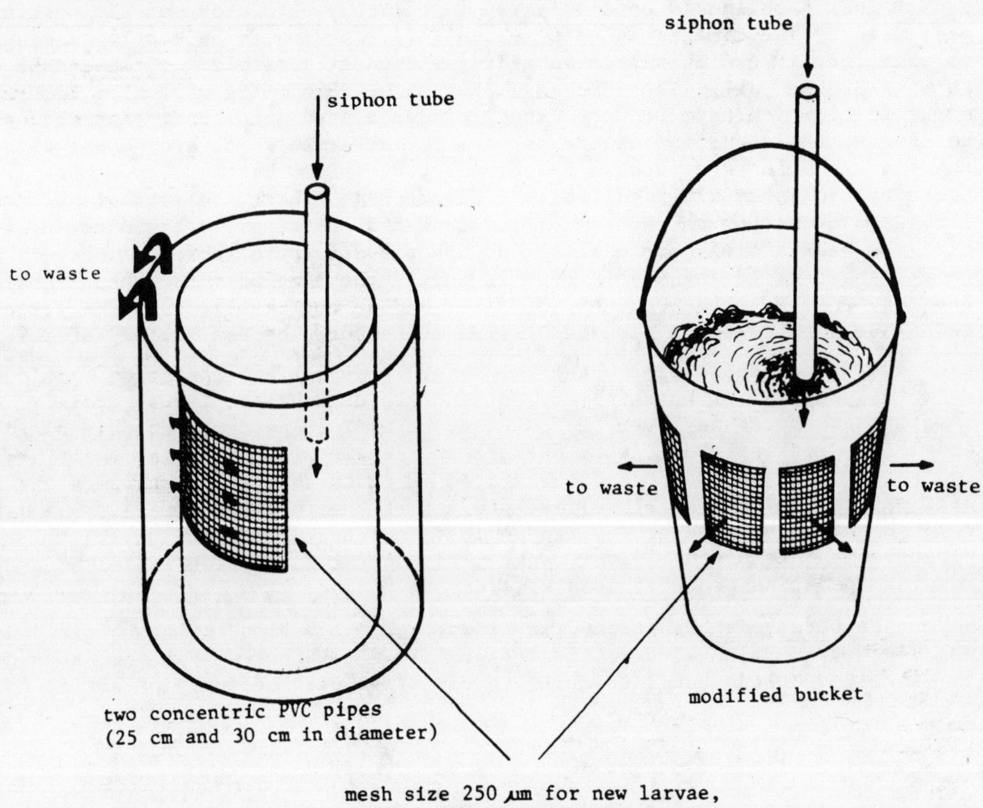

Figure 8. Larval tank cleaning

or the animals appear in poor condition[1/], or measurement shows a low DO_2 level, the water <u>must immediately be totally exchanged</u>. This is done by operating the turn-down drain until the water depth is only about 10 cm, flushing the tank with "new" water for 10-15 min and then filling up to 70 cm again. The "new" water used for flushing and replacement must be pre-aerated, 12°/oo salinity and the same temperature as the 'old' water and the tank aeration system must be kept on throughout the operation;

(f) some hatcheries regularly treat the larval water with antibiotics as a prophylactic (Manual Section 5.7.3), but this technique is not recommended at these larval densities;

(g) some freshwater prawn and marine shrimp hatcheries maintain a level of 10 ppm of the sodium salt of ethylene diaminotetraacetic acid (EDTA) in the larval rearing water, believing it to improve productivity. The exact mechanism of the improvement caused by this chelator is not well known.

5.2.5 Light

Exposure to direct sunlight appears to be harmful to the larvae, especially in a "clearwater" rearing system. It is essential, however, to make some light available to the larval tank. This should be sunlight or light with the same spectral quality. It is therefore recommended that 90 percent of the tank surface be covered. The material used to cover the tank can be whatever is locally and cheaply available, provided it does not disintegrate when exposed to sunlight, heavy rain or strong winds.

5.2.6 Hygiene

General water treatment is dealt with in Manual Section 4.4. In addition, it is good practice (though seldom done, because of the time and money it takes) never to use portable equipment for more than one tank. Thus each tank should have its own nets, siphon tubes, spare filters, etc. Water should never be moved from one larval tank to another. Submersible pumps are often used for water transfer but should never be used in the larval tanks themselves since they are a potential source of disease transfer, too. Larval tanks should always be drained by gravity or siphon and the submersible pumps only used in water storage or mixing tanks.

Some hatcheries find the sterilization of all their equipment in a solution of potassium permanganate at pH 3 between each larval cycle is a valuable hygienic practice.

Between larval rearing cycles, tanks should be routinely disinfected. Failure to do this usually results in massive blooms of organisms, such as *Zoothamnium*, *Epistylis*, hydroids, etc., which are harmful to the larvae. Disinfection does not eradicate these organisms but does effectively control their growth. Disinfection can be done by a number of techniques:

(a) scrape the tank, treat with 1.5 ppm chlorine (60 ppm 'chlorox' solution or 6 ppm commercial bleach powder (Manual Section 4.4(b)) for one day, flush/rinse, dry in sunlight for one day, and rinse/flush again before use;

(b) scrape the tank, spray with 250 ppm formalin solution, expose to sunlight for one day, flush/rinse and re-use.

5.3 <u>Feeding</u>

A wide variety of feeds are employed by different hatcheries, including the nauplii of brine shrimp (*Artemia salina*), *Moina* spp., fish eggs, squid flesh, frozen adult *Artemia*, flaked adult *Artemia*, fish flesh, egg custard, worms and compounded feeds. This manual describes only one feeding regime, which the authors have found effective. It is recommended that the readers also experiment with other materials locally available.

1/ Larvae in "poor" condition are sluggish, not active; do not appear strong enough to swim against the air bubbles; do not respond well to feed; are only at the edges of the tank; and sometimes jump out of the water (see also Manual Section 5.4). Non-feeding larvae are noticeable by the lack of the normal brown colour which is caused through the consumption of brine shrimp nauplii

Two feeds are employed, namely brine shrimp nauplii (hereafter called BSN) and prepared feed (hereafter called PF). Methods for preparing these feeds before use are given in Appendixes 3 (BSN) and 4 (PF). BSN are small crustacean nauplii hatched from cysts which can be bought in cans. The PF referred to in this manual is an egg/mussel mixture.

Most freshwater prawn larvae do not feed on the first day (hatching day). However, some do feed and it is good practice to provide some BSN even on this first day. From then until the fifth day, BSN are fed twice per day, in the morning and in the evening. As with all larval feeding, the amount of BSN given depends on visual examination of the larval water. Freshwater prawn larvae do not actively search for food, which is why BSN (which swim actively in the same part of the water table as the larvae) are so valuable a feed. The idea is therefore to always have BSN present in the tanks in sufficient numbers for the larvae to "bump" against. The amount of BSN required at any one time depends primarily on the tank volume, not on the number of prawn larvae present, although the latter of course controls the rate at which BSN are consumed. (This concept can be clearly illustrated by the following example: suppose a freshwater prawn larva will consume 50 BSN/day. In this case, if you have 150 000 larvae in a tank, you will need to provide 7.5 million BSN/day. However, as an extreme case, suppose you only have one larva in the tank: if you put in only 50 BSN, will the larva find them? Thus it is the density of the BSN that counts, not the total quantity.)

As a guide, there should be about 1-5 BSN/ml directly after feeding, depending on the age of the prawn larvae, and 1 BSN/ml left in the water just before the next BSN feeding time (Figure 7a). If there is more than 1 BSN/ml at the latter time then you have been overfeeding or the larvae are not feeding well; if there is less than 1 BSN/ml, you should add more this time than last time. A density of 1-5 BSN/ml in a tank with 10 m^3 of water means that 10-50 million BSN have to be added. The quantity of brine shrimp cysts ("eggs") necessary to produce 1 million BSN depends on the source of brine shrimp cysts used and the preparative treatment they are given; it is usually stated on the cans. As a rough guide, however, you can assume that 50-250 g of *Artemia* cysts will be required to produce the 10-50 million BSN required for the daily feeding of 10 m^3 larval tank. Normally one larval cycle in this size of tank 1/ will consume 2.5-5 kg of brine shrimp eggs.

Three days after hatching, PF can be introduced in increasing weaning quantities. By day five BSN should be given only at the evening feeding to ensure the presence of food at all times. Use of PF for the night feeding is not recommended because the quantity necessary to supply the requirement throughout the night in one feeding would foul the water. By day five PF should be given as 4-5 meals spread throughout daylight hours at 1.5-2 hour intervals. Once again, the exact quantity of food to be given at each meal cannot be prescribed. It depends on the utilization of the feed by the larvae and must be judged visually by the operator.

The quantity of feed necessary will obviously rise as the larvae grow. As a rough guide, you should expect to use 12-16 kg of PF for each larval cycle per 10 m^3 tanks 1/. Initial quantities of PF at day five would be about 30-60 g/tank at each feed and will rise to around 200 g/tank/feed.

The basic rule is that each larva should be seen to be carrying a particle of PF immediately after a PF feeding. Underfeeding will lead to starvation, cannibalism and slow growth; overfeeding (especially if large quantities of PF are obvious before the next feeding time commences) will cause water pollution. Pollution through overfeeding is obvious through the presence of PF particles before the next feeding or if there is a severe "foam" or "scum" on the water surface. Should water pollution occur by error, the water must be immediately exchanged (Manual Section 5.2.4(e)).

1/ Approximate stocking density, 30-50/1; expected post-larval production, 10-20/1

PF which is about 0.3 mm in size should be used up to day ten; from then to metamorphosis PF of 0.3-1.0 mm in size should be used. The particles of PF must be kept close to the larvae; this is an additional reason for ensuring vigorous aeration in larval tanks.

A feeding schedule is given in Table 4.

5.4 Growth Rate and Metamorphosis

In practice it is not necessary or feasible to monitor the progress of the larvae by microscopic examination. However, a key to the various larval stages is provided in Appendix 5 for those who wish to do so. The hatchery operator very quickly becomes able to gauge whether his larvae are feeding and growing well by crude visual examination, using a white sight board (Figure 7b) and observing the behaviour of the larvae. Healthy larvae swarm at the surface of the water (especially in the first 10 days), feed actively, have a red-brownish pigmentation, and are not observed to cannibalize. Unhealthy larvae accumulate at the bottom of the tank and are often bluish in colour. Food consumption drops and, if the problem is already severe, dead larvae are observed. Healthy larvae swim tail first, head down and ventral side up.

The time taken for a larval batch to metamorphose varies according to feeding and environmental conditions, particularly temperature. In a healthy, well fed batch, which is maintained within the optimum temperature range (Manual Section 5.2.2), the first few post-larvae should be observed about day 16-18. Most of the larvae should have metamorphosed into post-larvae by day 25-28 and it is seldom economic to maintain the batch longer than this in order to wait for the last few larvae to metamorphose.

Metamorphosis to post-larvae is characterized by a radical change in behaviour and appearance. For the first time the animals resemble miniature adult prawns and, instead of swimming freely in the water, many crawl or cling to the tank surfaces. Once the majority (90-95%) of the larvae have metamorphosed into post-larvae they should be harvested from the larval tank and transferred to holding tanks. At this time, brackish water is no longer required and the post-larvae can be maintained in fresh water alone.

5.5 Harvesting and Holding Post-Larvae

Although post-larvae can withstand the physiological shock of sudden transfer from $12^{o}/oo$ water into fresh water, it is not normally recommended to harvest them from brackish water and transfer them directly into holding tanks containing fresh water. In practice the animals are best acclimatized to fresh water in the larval tank. Once the majority of larvae have metamorphosed (at least by day 28) the tanks should be drained by means of the turn-down drain to a level of about 35 cm and gradually flushed with fresh water over a period of 2-3 hours. Aeration, as always, must be continued during flushing. The post-larvae can then be harvested and transferred, or the larval tanks refilled to 70 cm with fresh water and the animals retained in them. If the latter is done, the post-larvae should only remain in the larval tanks for a few more days, with frequent water exchange before transfer to a larger holding tank or the biomass will become excessive and water quality deterioration and cannibalism will occur.

Post-larvae can best be harvested from the larval tanks by reducing the water level and using dip nets. This is facilitated by covering most of the tank and allowing the post-larvae to concentrate in the illuminated area. The remaining post-larvae can be obtained from the tank by removing the filter sock and flushing them to the exterior, where they can be caught by net, by means of the turn-down drain. Care must be taken to see that the animals do not become stranded, or oxygen levels depleted during this procedure.

For stock record purposes it is necessary to estimate (Appendix 6) the number of post-larvae obtained from each cycle. This should be between 100 000 and 200 000 for the 10-m^3 tank described in this manual.

The post-larvae can be transferred to the stock or holding tanks, which contain pre-aerated fresh water, in any suitable container. These should not be overcrowded with post-larvae, nor left too long before transfer, or oxygen depletion will occur. Aeration or packing (Manual Section 5.6) should not be necessary unless the holding tanks are on another site.

Regarding size, 50-m^3 concrete tanks are convenient for holding post-larvae prior to transport for stocking in ponds. Normally, post-larvae are retained in these tanks for 1-4 weeks prior to stocking. The length of time they stay in the holding tanks depends on the demand for post-larvae at that time. Whilst in the holding tank the post-larvae must continue to have their rearing water exchanged (equivalent to 200 percent per week) and aerated. Densities of up to 5 000 post-larvae/m^2 can be successfully maintained for one week (note that once animals become post-larvae, it is normal to refer to density on an aerea, per m^2, rather than a volume, m^3 basis), or up to 2 000/m^2 for one month under these conditions. If held for one month, survival is enhanced if density can be reduced to 1 000/m^2. Providing increased surface area in the tank by suspending mesh sheets from floats is good practice.

There is no need to continue feeding BSN after metamorphosis and the post-larvae can be weaned directly on to the diets utilized for pond feeding (Manual Section 8.3.1). However, the authors have found it convenient to utilize a floating diet in the holding tanks. By this means the quantity to feed, which depends once more on demand, can be visually gauged much more easily. The young post-larvae, although they tend more and more to cling to and crawl on surfaces, still swim quite actively in the water and utilize a floating diet well. A floating catfish diet or even an expanded dog feed is adequate.

Some farms will want to stock production ponds with juveniles instead of post-larvae. If they do not want to have nursery ponds themselves, the hatchery will have to have this facility on site. The topic of nursery ponds is dealt with in Manual Section 8.8.1.

5.6 <u>Transporting Post-Larvae</u>

Cooled and aerated fish transport tanks would be ideal for transporting freshwater prawn larvae from the hatchery holding tanks to the pond site but they are rarely available or affordable. For distances up to one hour to the pond site aerated garbage cans can be used for transport. A 100-litre trash can, holding 40 l of water, will hold 30 000 post-larvae. Baffles should be inserted in the can to prevent excessive water movement during transport.

For longer distances the same technique as is used for transporting aquarium fish can be utilized, namely transport in plastic bags containing 1/3 water and 2/3 air or oxygen (Figure 9). A transport rate of 125-250 post-larvae/litre is practical. A 45 x 80 cm bag holding 8 litres of water will take 1 000-2 000 post-larvae. The corners should be rounded off with rubber bands to prevent animals getting trapped there. The top is twisted, bent over, and sealed tightly with a rubber band after the bag has been inflated with air or oxygen.

These inflated bags can be used to transport post-larvae very long distances (up to at least 16 hours travelling time by road). When inserted into insulated "styrofoam" boxes they can be used to ship post-larvae by air most effectively. Enclosed in non-insulated boxes they can be used for night (cool) journeys by rail, for example. For long day-time (hot) journeys, these plastic bags can be stacked on shelves in a home-made transport box mounted on a truck.

The transport box (Figure 10) is insulated and garbage cans filled with ice keep the temperature down. Temperature within the box is maintained steady through the use of battery-operated fans. Very good transport survival can be achieved in this way.

In addition to reducing metabolic activity during transport by lowering temperature, it is important to use water from the holding tank to fill the plastic transport bags.

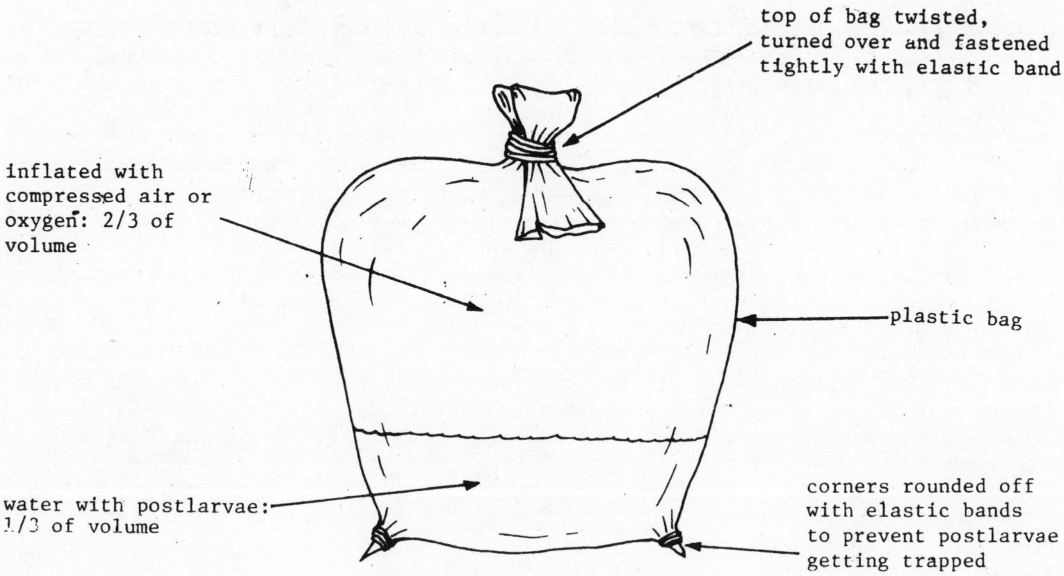

Figure 9. Postlarval transport bag. Two bags, one inside the other, are safer than one! Bag size is not important, but a 45 cm x 80 cm bag will hold 1 000 - 2 000 postlarvae in 8 litres of water (and 16 litres of air or oxygen)

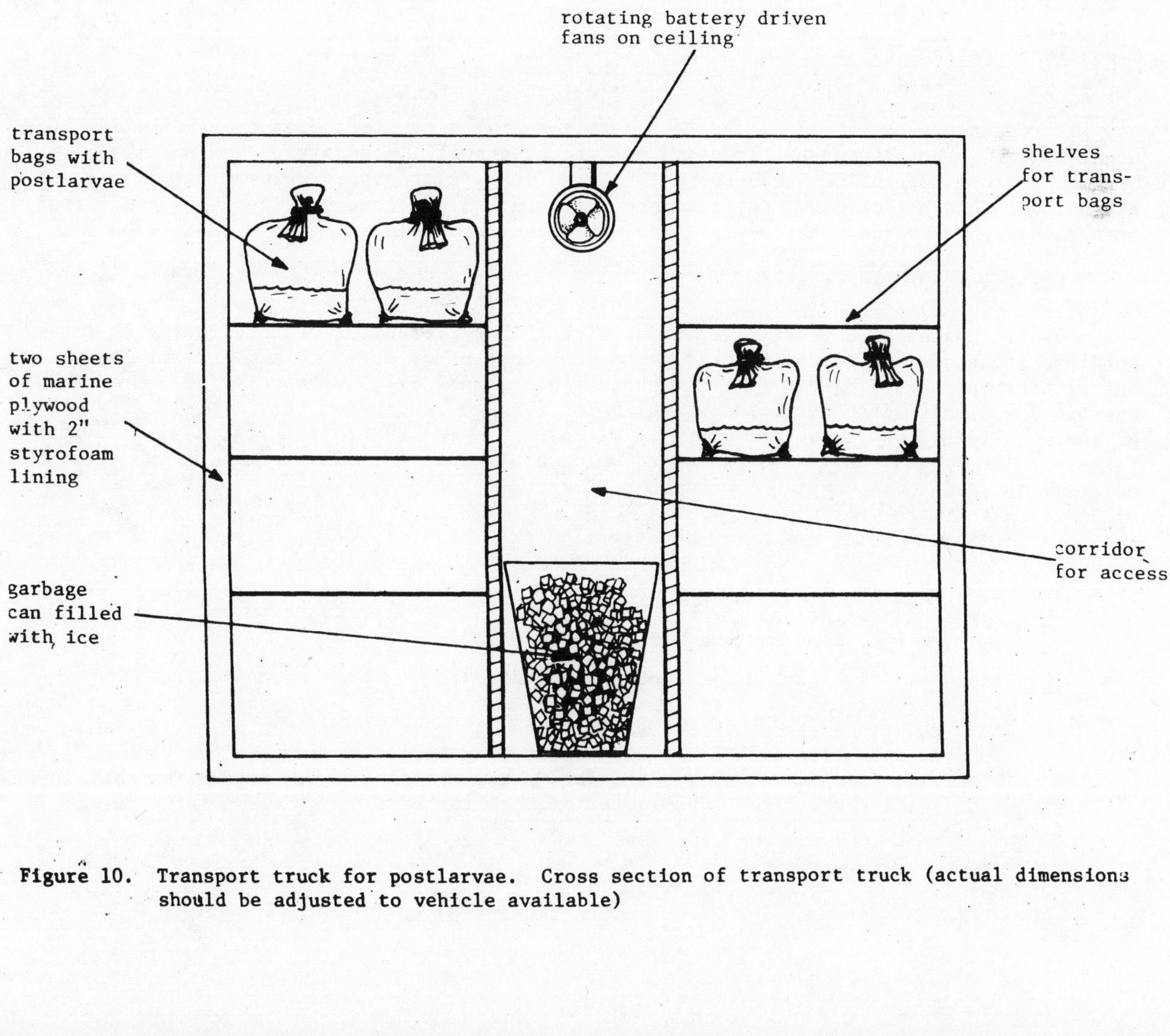

Figure 10. Transport truck for postlarvae. Cross section of transport truck (actual dimensions should be adjusted to vehicle available)

If the post-larvae are placed into "new" water for transport, many will moult during the journey with consequent losses through cannibalism. Some hatcheries, however, add a little sea water to the transport bags, claiming that survival rates are better in brackish water than in fresh water.

To facilitate stocking at the pond site it is normal to standardize the number of prawns in each transport bag. It is therefore necessary to estimate the quantity of post-larvae as accurately as possible when the bags are filled (Appendix 6). Exact accuracy is impractical and the value of accuracy must be weighed against the losses of post-larvae which will be caused by excessive handling. However, estimates must be reasonably accurate because it is upon this figure that pond stocking and feeding rates will be based and charges for post-larvae levied.

The survival rate of 7-day old (after metamorphosis) post-larvae during shipping is much higher than 1-day old animals. It is not good practice to ship post-larvae of widely different age groups although, because of the method of larval culture, the post-larval age will vary by a day or two.

5.7 Problems

5.7.1 Management

Most hatchery problems are caused by poor management. The commonest causes of larval loss are not actually mortalities due to poor water quality or disease, but physical losses due to simple operator error during tank cleaning and siphoning, water exchange, etc. Poor internal (i.e., within the hatchery) water quality is inevitably caused by inadequate water exchange, poor daily observation of the larvae, overfeeding, total failure of the aeration equipment, or individual tank aerators being blocked or left turned off, for example.

Late larvae (soon before metamorphosis) jump quite a lot. Some hatcheries glue mosquito screening, 12 cm wide, at the water surface of the tank which helps to prevent larvae becoming stranded.

It cannot be overstressed that while this manual attempts to lay down guidelines for a particular method of freshwater prawn culture, successful hatchery operation is a blend between factory discipline and husbandry. If a close relationship between the farmer and his animals does not develop, the hatchery will fail. The hatchery manager and his staff must always observe the behaviour of his larvae and the condition of the tanks closely.

5.7.2 Disease and predation

Several diseases have been reported which affect freshwater prawn larvae. It is beyond the scope of this manual to provide a complete list of these or a means for their identification and the reader is referred to Sindermann (1977), Johnson (1977), Goodwin and Hanson (1975) and Hanson and Goodwin (1977) for further reading on this topic. Those hatchery operators who encounter possible disease problems should seek the advice of local fishery pathologists (where available) and microbiologists for identification and treatment purposes.

The following brief notes may, however, help:

(a) a disease problem is usually secondary to or aided by a primary failure in tank hygiene, insufficient water exchange, feed quality or quantity and low dissolved oxygen level, which result in bad larval condition;

(b) protozoa are a common cause of larval "disease". These commonly include the genera *Epistylis* and *Zoothamnium* and, less commonly, *Vorticella*. These protozoa move about and attach themselves to the body surface and the gills of the larvae. They are normally cast during the moulting process but can seriously affect larval movement, feeding and gill operation. They are also often evident on tank surfaces. Ciliates feed on bacteria and the link with poor tank maintenance is obvious;

(c) the medusan stage of small hydrozoans has been reported to actively prey both on brine shrimp nauplii and freshwater shrimp larvae. Problems with hydrozoans are particularly acute when surface water sources are used. The importance of using ground water and, if this is unavailable, proper water treatment (Manual Section 4.4.2) is therefore emphasized;

(d) bacterial infections take two forms: (i) chitinolytic bacteria which erode the surface of the exoskeleton, often following physical damage, and appear as black spots or lesions or can cause loss of appendages, and (ii) filamentous bacteria which can clog the gills and interfere with respiration;

(e) viruses have not yet been demonstrated in freshwater prawns;

(f) the muscles of larvae subjected to stress or low oxygen levels sometimes become opaque or whitened and often recover if the exterior problem is cured. The syndrome has only been associated with excess sunlight (Manual Section 5.2.5);

(g) fungal infections of larvae have been observed but are often eliminated by better food hygiene and a reduction in larval density;

(h) formalin (200-250 ppm daily dip for 30 min)/ malachite green (0.2 ppm daily dip for 30 min) and copper sulphate (0.4 ppm dip for 6 hours) have sometimes proved an effective remedy for the disease syndrome noted in (b), (c) and (g) above. Where the treatment period is short it is best to apply it when the tank water level is very low (10-15 cm) so that it can be rapidly flushed with "new" $12^{o}/oo$ water after treatment. The flushing process should continue for one hour. Aeration must continue during treatment as normal. Formalin can also be used at a lower level of 25-30 ppm for a longer period, followed by a water change after 24 hours. Mortalities from hydroid infestation can also be reduced by transferring healthy larvae to newly disinfected tanks every 5-10 days. Antibiotics are sometimes used to control filamentous bacteria (*Leucothrix* spp.) and as a prophylactic (Manual Section 5.7.3);

(i) as noted in (a) above, a severe disease problem is normally initiated by a more basic management failure. Experience shows that it is better to discard any seriously infected batch of larvae rather than try to treat and save the weakened survivors. Do not waste time on treatment, try to find the basic cause and eliminate it, discard the prawns, drain the tank and disinfect it (Manual Section 5.2.6). Dead larvae are eaten by live ones; if you see mortalities, the problem is already severe.

5.7.3 Prophylaxis

While, as mentioned in Manual Section 5.7.2, treatment of diseased larvae is often useless, some hatcheries utilize a prophylactic regime employing antibiotics and/or sulpha drugs. For example, streptomycin and bipenicillin are added to the larval rearing water at levels between 1.25 and 2.5 ppm every 2 or 3 days during rearing in the conical tank system for larval culture (Manual Section 5.8.3). Those levels were increased to 5 ppm when necrosis or increases in filamentous bacteria, followed by mortality were observed. The treatment is given when the tank water level is at its lowest just before refilling commences. Another hatchery alternates the use of 1-2 ppm of either Ampicillin (synthetic penicillin) or Oxytetracycline every three days commencing immediately after hatching.

While these treatments seem to pay off, their regular use cannot be recommended because of the inherent danger of producing resistant strains of bacteria. The cost of using antibiotics in large volumes of water is also high.

5.8 Alternative Hatchery Techniques

The fact that there are many alternative techniques used for culturing freshwater prawn larvae other than that detailed in this manual has been referred to a number of times before. Some of these techniques are mentioned below.

5.8.1 High-density culture

Economies in the consumption of water, food and the number of tanks required in the technique described in this manual can be effected by stocking larvae much more densely for the first 10 days (60-100/ml) before dividing them out into two or three similar sized tanks after one week for completion of the larval cycle. In a large hatchery this technique, together with careful programming of larval cycles, can increase the potential productivity substantially. (See also Manual Section 5.8.3)

5.8.2 "Green water" culture [1/]

This system has been successfully used in Hawaii to control blooms of organisms harmful to freshwater prawn larvae and has been claimed to act as a buffer against ammonia build up. For some reason this technique has had limited success outside Hawaii and is not now in use in Thailand, where the "clear water" technique described in this manual is favoured.

"Green water" is a mixed phytoplankton culture in which *Chlorella* spp. is dominant. Its cell density is about 750 000-1 500 000 cells/ml. A fertilizer solution in tap water is added to the tank at least once per week to maintain the culture. This solution provides a mixture of 4 parts of urea to 1 part of NPK (15:15:15) garden fertilizer applied at the rate of 185 g/10 m^3 of water. Tilapia, *Sarotherodon mossambicus* are held in the tank at the rate of about 1:400 litres to graze on and control filamentous algae. Copper sulphate, at a rate of 0.6 ppm is added to the green water tank once per week to control rotifers. The tilapia also help to fertilize the culture. Ten ppm of the sodium salt of ethylenediaminotetraacetic acid (EDTA) is also sometimes included in the "green water" culture. The green water is prepared at the same salinity as the larval rearing water ("green water" does not thrive at more than 12°/oo salinity) and is used as replacement water during exchange procedures instead of plain brackish water. A green water culture is never used for larvae if the culture is more than three days old. Part must be discarded or used for filling larval tanks and the rest diluted regularly to avoid phytoplankton "crashes" (with the ensuing DO_2 problems) occurring in the larval tanks.

5.8.3 Conical tank rearing

An intensive technique for freshwater prawn larval culture which produces over 50 post-larvae/litre has been developed by Aquacop (1977; 1979a). This technique has now been adopted by an experimental hatchery in Indonesia, which recently (Suharto *et al.*, 1982) reported post-larval production rates between 60 and 110/litre. The technique is sophisticated and includes the use of conical fibreglass tanks, clear water, total daily exchange of water, temperature control, routine antibiotic administration and water chlorination.

5.8.4 Recirculation systems

Recirculation systems involving the use of biological filtration are being developed to conserve water and energy usage. These range from simple systems utilizable by the small hatchery (Singholka and Sukapunt, 1982) through experimental systems involving the use of ozone (Menasveta, 1982) to that in use in a large freshwater prawn hatchery in Honduras (Wulff, 1982).

The latter technique involves a water exchange of 6 times the larval rearing tank volume per day through a biological filter with no added water other than the replacement of evaporative losses. Before use the hatchery water is chlorinated, temperature controlled, and filtered to 8 μm. Post-larval production of 25-35/litre in less than 4 weeks is routine.

In its simplest form this technique (which reduces the water consumption of a hatchery enormously and makes inland hatcheries possible) consists of recirculating the larval rearing water through a graded sand/gravel filter (similar to that described in Appendix 2). The water can be recirculated by mechanical (See Manual Section 4.5) or air-lift pumps.

1/ This section was prepared partly through reference to Fujimura (1978)

5.8.5 Non-coastal hatcheries

Though the ideal hatchery site for freshwater prawns is coastal (Manual Section 3.1) many farmers establish 'backyard' hatcheries on inland sites (Singholka, 1978) where seawater supplies must be trucked at considerable expense. It is hoped that the development of simple recirculation systems (Manual Section 5.8.4) will help to reduce the costs of such hatcheries.

Artificial sea water has also been used in hatcheries, but is expensive. Recently a report of preliminary experiments in the use of the residues from salt evaporation pans has appeared (Tunsutapanich, 1980a). This opens the possibility of establishing freshwater prawn hatcheries in inland as well as coastal salt pan areas.

6. REARING SITE REQUIREMENTS

6.1 Market

The prime requisite for the success of any farm is that there is a market for its produce. This is no less true for freshwater prawn farms than for those for other aquatic or terrestrial organisms. The scale, nature and locality of the market is the first topic that should be considered in the selection of a site and the results of this evaluation will determine the manner in which the farm is designed and operated. Despite the obviousness of the above statement it is surprising how often the market is the last criterion to be investigated.

Further reading on this very site-specific topic may be found in an interesting report by Shang *et al.* (1980) applicable to Thailand and another by New *et al.* (1977).

6.2 Water

Fresh water is normally used for rearing freshwater prawns from post-larvae to market size, though the successful experimental use of partially saline water has been reported. Tidal water fluctuating between 12°/oo and 25°/oo has been utilized in Western Samoa (Popper and Davidson, 1982), while salinities of at least up to 10 °/oo gave as good results as freshwater in South Carolina, USA (Smith *et al.*, 1982).

Barnes (1982) has also reported the use of brackish water for freshwater prawn culture in Israel. Clearly, therefore, freshwater prawn farming need not necessarily be restricted to sites with freshwater supplies in the future. Since the results mentioned above are from research rather than commercial units, some caution should be applied to the use of brackish-water sites for farms, for the moment.

As with any other type of aquaculture, the quality and supply reliability of water at the site is a critical factor in site choice. As for freshwater prawn hatchery water supplies (Manual Section 4.4), little is known about the optimum characteristics of water supplies for rearing sites. Again, there are indications that growth rate is much less in hard water and it is advisable to site farms where the water has a hardness less than 150 ppm (preferably less than 100 ppm)[1/]. Provisional criteria for water supplies for freshwater prawn farming are as follows:

pH	:	7.0 -8.5
Total hardness	:	<150 (preferably <100) ppm (as $CaCO_3$) and >40 ppm
Temperature	:	18°-34°C (optimum 29°-31°C)
Dissolved oxygen	:	>75% saturation

Temperature below 14°C or above 35°C usually prove lethal to freshwater prawns.

1/ Reported as ppm $CaCO_3$. Recent information (Bartlett and Enkerlin, 1983) indicates that high calcium levels may not be detrimental unless combined with high carbonate levels. Successful freshwater prawn rearing was reported with the use of 940-1060 ppm total hardness where the CO_3 content was nil but SO_4 and Cl were high.

In addition, the water supply must be free from pollution, particularly agricultural chemicals, and as predator free as possible. This may be achieved by screening (Figure 19) or by the use of well water.

Underground water, because of its chemical and microbiological quality and its lack of predators, is undoubtedly the preferred water source for freshwater prawn farming. However, the lack of this source does not eliminate the use of sites only having access to surface water supplies (rivers, lakes, reservoirs, irrigation canals, etc.). Such sources may be successfully employed but the farmer must be aware of the extra risk that their use brings. Screening of the water supply helps to reduce initial entry of predators but cannot clean up chemically polluted water or water containing disease organisms. The location of other existing or planned freshwater prawn farms should be taken into account so that the risk of contamination of the water supplies of one farm by the effluent from another can be assessed. If surface water is to be used, farms close to a waterfall bringing water from a remote and unpolluted watershed or below the dam of a reservoir (though such water, if driven from the epilimnion, may initially be high in hydrogen sulphide) are well placed.

Water for the farm is required for four major purposes:

(a) filling ponds;

(b) compensation of seepage and evaporation losses;

(c) "flow through" (for "continuous culture" system only);

(d) emergency (or "flushing") use.

While the minimum farm size for economic viability may depend on other factors, the quantity and continuity of water supply available sets an absolute technical limit on the pond area and on the potential productivity of the farm. To fill a 0.2 ha pond of average water depth 0.9 m, requires 1 800 m^3 of water. Since it is desirable to be able to fill the pond within 12 hours it follows that it must be possible to extract up to 2.5 m^3 (2 500 litres) per minute from the water source for this pond. Normally it is seldom necessary to completely fill a drained pond. In "continuous" 1/ culture it may only be necessary once every two or three years while in "batch"1/ culture it may be necessary once every 6-8 months. However, there will be occasions when, because of pond water quality, it is necessary to flush the pond and refill it while prawns are growing in it. Thus, the water source must be available at that volume throughout the growing period. It is very unlikely that it will be necessary to fill more than one pond, in a small farm, at the same time however, so five 0.2 ha ponds will not require a maximum water supply five times larger than one 0.2 ha pond. A guide to the maximum water consumption for different sized farms is given in Table 5.

When determining the amount of water available on a specific site for freshwater prawn farming the rainfall pattern must be taken into account. This may be sufficient to replace or exceed evaporative and seepage losses, at least at some time during the year.

In addition to filling the ponds, it is, at the very minimum, necessary to have enough water available throughout the growing period to replace evaporative and seepage losses. Evaporative losses depend on solar radiation, wind and relative humidity and are governed by the climatic features of the site. Seepage losses depend on soil characteristics of the farm area, mainly its permeability. Seepage losses may be minimal where the water table is high or where the water level of the pond is the same as in adjoining fields (e.g., in a paddy field area). In other cases, however, particularly where pond construction is poor, seepage losses can be very great. The quantity of water necessary for this purpose must be assessed locally and the cost of providing it is an important economic factor. As ponds mature, ponds tend to "seal" themselves, through the accumulation of detritus and algal growth, thus limiting seepage losses.

If the water supply available exceeds the farms' requirements for pond filling and the replacement of evaporative and seepage losses throughout the year, "continuous" culture may be practised if desired. When the water supply is only seasonally sufficient, "batch"

1/ For definitions of these terms see Appendix 10

culture will have to be practised and the timing of stocking and harvesting procedures carefully selected. Greater productivity [1/] can be achieved where water is continuously running through the pond, though the flow must not be so great as to significantly remove nutrients from it.

Normally, except in cases of severe evaporative and seepage losses a regular flow rate of 0.14-0.28 m^3 (140-280 litres) of water/ha/min is sufficient to cover water losses, while a flow rate of 0.56 m^3 (560 litres)/ha/min is usually enough for the "continuous" culture system of management. Additionally it is necessary to have water available for flushing the ponds if there is a water quality problem. This contingency has been taken into account in the calculations of water requirements for freshwater prawn farming given in Table 5.

6.3 Power

A source of electricity is desirable but not essential for a freshwater prawn farm. A variety of power sources are used for supplying the energy necessary for water movement on the farm including: water power itself (gravity and current flow), wind, electricity, petrol and diesel fuel and wood.

Electricity is desirable, although it need not be the sole source of energy, for powering lights, wells and feed-making equipment. Ideally, all water flow is by gravity once the water enters the farm site but this depends of the nature of the site (Manual Section 6.4). In practice, most farms use electric or fuel-driven pumps for supplying water to the ponds and many also use them for draining the ponds during harvesting or water exchange.

The most suitable power source to use is entirely site specific and depend upon such factors as equipment availability, unit power costs and the characteristics of the site and its water supply. No single method can therefore be recommended as best in this manual.

Some small farms prepare cooked feed using wood as a fuel source, while others utilize the time-old methods of wind and water power for transporting water. One interesting example of the diversity of power sources utilized in Thailand is provided by three adjacent farms, all drawing water from the same river. One uses a diesel engine pump, another a windmill (Figure 11) and the third a water-wheel driven by the current of the river (Figure 12).

6.4 Topography and Soil

Standard preliminary considerations in the selection of any site for inland aquaculture include:

(a) a study of meteorological records to determine the amount and seasonality of rainfall, evaporation, wind speed and direction and relative humidity;

(b) a topographical survey, including transects, to evaluate slope and to determine the most economic ways of constructing ponds and moving earth;

(c) taking soil cores up to 1 m deeper than the expected pond depth for soil analysis, including classification and soil chemistry.

The ideal site slopes gently (not more than 2 percent) and ponds constructed on it can be gravity filled (either naturally or by creation of a dam) and drained. In practice many successful farms exist where the only feasible method, not only of filling but also of draining the ponds is by pump. The cost of pond filling and drainage operations, which are governed by the characteristics of the site, must be considered before the site is chosen.

There must be enough soil for pond construction, either by excavation or above ground bunds erection. If there are rocks and boulders and tree stumps present, it must be economic to remove them in order to make the land bottom flat. The site should preferably not be so constructed that irregular shaped ponds are needed. Rectangular ponds are more efficient to operate.

1/ What that level of productivity is, depends on many other factors, as will be explained later

Figure 11. Freshwater prawn farm with windmill powered water supply

Figure 12. Water wheel used to supply water to freshwater prawn farm

The farm should not be sited in an area which is subjected to severe periodic natural catastrophes, such as floods, typhoons, land-slips, etc. If it is decided, for other reasons, to site a farm in an area subject to floods, individual pond bunds should be constructed so that they are higher than the highest known water level at that site or the whole farm should be protected by a peripheral bund.

Apart from the dangers of water-supply contamination (Manual Section 6.2), the farm should not be sited in an area where the ponds themselves are likely to be affected by aerial drift of agricultural sprays; prevailing wind direction should therefore be taken into account. Siting ponds adjacent to areas where aerial application of herbicides or pesticides is practised is also undesirable.

Although supplemental food is given to freshwater prawns reared in earthen ponds (Manual Section 8.3), a considerable amount of their food intake is from natural sources (Stall, 1979; Weidenbach, 1982). It is therefore important to site the farm where the soil is fertile. Since a water pH of 7.0-8.5 is required for successful freshwater prawn culture it is preferable not to build the farm on potentially acid sulfate soils. These soils have pH values of 4.5 or less, together with high concentrations of soluble iron, manganese and aluminium and can be found far away from mangrove areas with which they are often identified. Despite their unsuitability, fish farms are frequently constructed on such soils.

Freshwater prawn ponds should be constructed on soil which has good water retention characteristics or where suitable materials can be economically brought onto the site to improve water retention. The water retention characteristics of soil are highly site-specific and the prospective farmer must seek the professional advice of soil engineers and fishery officials from local government departments including the Ministry of Agriculture and the Public Works Department.

If there are other fish farms or irrigation reservoirs in the area, the best advice is the experience of his neighbours. Pervious soils, which are very sandy or a mixture of gravel and sand, are unsuitable unless the water table is high and surrounding areas are always water logged. Soils which consist of silt or clay or a mixture of these with a small proportion of sand normally have good water retention characteristics. The clay content should not exceed 60 percent. Higher clay content soils crack during the dry season and necessitate repairs. Peaty soils are not suitable.

6.5 Access

6.5.1 Market

The important topic of the market for freshwater prawns and its relationship with site selection has already been mentioned (Manual Section 6.1). It is stressed again here because the farm must not only be close to its market but the road access must be good. It is essential that heavy trucks be able to reach the farm during harvesting and be able to take the animals away from the site without delay.

6.5.2 Fry supply

As mentioned earlier (Manual Section 5.6), there is no fundamental technical difficulty in transporting post-larval freshwater prawns long distances by road, rail or even air. However, there has to be a means of getting close to the pond site with a vehicle for stocking purposes. It is not satisfactory to bring post-larvae long distances only to have further local delays due to poor local access. Most important of all in selecting the site of the farm, the cost of obtaining the post-larvae for stocking purposes must be included in the economic assessment. Transport costs can add enormously to basic stocking costs and prices of the post-larvae themselves tend to rise as the distance between the farm and the nearest hatchery increase (and thus the competition between hatchery operators decreases).

6.5.3 Feeds

The availability and cost of getting feeds to the farm site must also be considered; a 40-ha farm, for example, may require an average of about 5 t of dry feed per week. A monthly delivery of (say) 20 t of feed also requires good access to the farm.

6.6 Sympathetic Authorities

Site selection must take into account the local and national government regulations concerning water usage and discharge, land use, movement of live animals, import of non-indigenous stocks (where *M. rosenbergii* is not already present), disease monitoring, taxation, etc. In most countries where freshwater prawn farming is technically and economically viable, these regulations are much less restrictive than those, for example, applying to the culture of temperate aquatic species in Europe and the USA. Most tropical governments are keen to encourage freshwater prawn farming, so, seek the advice of your local inland fisheries department. You will find their officers helpful and anxious to participate in your project.

Similarly the ease of access to these same individuals when the farm is in operation is an important factor in site selection. No matter how competent a prawn farmer you are (or become), there will come a time when you need help such as water analysis, disease diagnosis and technical advice. Those types of assistance can be obtained from government, university and private sources. Do not site your farm too far from someone who can heed your cries for "help". Speedy access to qualified personnel and to well-equipped laboratories is invaluable. Always keep in touch with your local fisheries officers but do not expect them to know <u>all</u> the answers. <u>No one</u> does, <u>not even the authors</u> of this manual.

6.7 Labour

Small freshwater prawn farms can be successfully maintained by unskilled labour but outside assistance from government or commercial sources (hatchery operators, feed supplies, etc.), is necessary at times of stocking or harvesting. Larger farms require a competent, on-site manager.

The amount of labour utilized on freshwater prawn farms varies considerably. One 40-ha farm in Thailand is run by two senior staff and six labourers. One man should be able to take care of the normal maintenance including feeding but excluding harvesting, of a 1-2 ha freshwater prawn farm.

7. FARM FACILITIES

Freshwater prawns are stocked into concrete and earthen reservoirs, ponds, irrigation ditches, cages, pens and into natural waters. Cage and pen culture is experimental, while the production from irrigation ditches is low; stocking into natural waters and reservoirs is not strictly prawn farming. This section of the manual and that dealing with farm operation will therefore deal only with ponds. Some notes on alternative methods are however given in Manual Section 8.8, together with some references for further reading.

A freshwater prawn farm is very similar to a freshwater fish farm. The authors do not provide a detailed farm design in this manual since every farm must be unique to its site characteristics. A photograph of a large freshwater prawn farm is given in Figure 45. This section of the manual does, however, discuss some general principles of farm construction and further reading on pond construction can be found in Sivalingam (1974) and Wheaton (1977).

7.1 The Pond

7.1.1 Shape and size

Rectangular ponds are most suitable for the type of harvesting (seining) usually practised in freshwater prawn farming. The maximum width should not exceed that through which a seine can be conveniently drawn from one end to the other. A convenient width is 30 m but widths up to 50 m are seined in Thailand. The length of the pond depends partly on the topography of the site and partly on the pond size and farm layout chosen.

The consensus of opinion is that the most easily managed pond size ranges between 0.2 ha and 1.6 ha, with most farms having ponds around 0.2 ha-0.6 ha. If kept to a 30 m width a 0.6-ha pond will be 200 m long. Narrow ponds should be oriented so that the prevailing wind blows down the long axis toward the drain end, to lessen the area of bund subject to wave erosion.

Large ponds are normally constructed wider than 30 m and often drained for harvesting. If draining is used for total harvesting, the size of the pond should be influenced by the maximum weight of prawns that the market will accept at one time without price deflation. For example, if a quantity greater than 300 kg of freshwater prawns would swamp the market and reduce prices it would be pointless to have a drainable pond greater than 0.15 ha in area [1/]. In practice, of course, wider ponds can also be seined but not so efficiently as narrow ones.

7.1.2 Depth

Average depth should be about 0.9 m, with a minimum of 0.75 m and a maximum of 1.2 m. Deeper ponds are difficult to manage and even ponds of the recommended depth may have to have part of the water drained or pumped out to facilitate seining operations at the deep end. Shallower ponds get too hot and support the growth of rooted aquatic plants.

The bottom of the pond must be smooth (Figure 13); there must not be projecting rocks or tree stumps in it which will prevent efficient seining and damage nets. It must slope gradually and smoothly from the water intake end toward the drain end so that, when drained, pockets of undrainable water in which prawns become stranded and die do not occur. A slope of 1:500 is suggested for ponds 0.4 ha or more in area and 1:200 for smaller ponds.

7.1.3 Bunds

Bunds must have a free board of at least 60 cm above the highest water level expected in the pond. They must also be high enough to protect the pond from exterior flooding. Proper compaction must be employed both in the construction of bunds and the treatment of the bottom of the ponds to maximize water retention. Where the retention characteristics of the soil on the site are not good, a core of impervious material brought from outside the site must be provided during bunds construction. This core should extend below the level of the bottom of the pond (Figure 14b).

The internal slope of the bund should preferably be 3:1 in sandy areas and never less than 2:1 (Figure 14a). The external slope should preferably be 2:1 and never less than 1.5:1. Properly constructed bunds are more expensive and use more land but failure to build them correctly may result in severe erosion (Figure 15). After construction they should be planted (Figure 16) with plants such as fast growing grass, *Phyla nodifera*, kudzu (a woody vine) or taro (dasheer), to help prevent erosion. The planting of large trees or plants with an extensive root system on top of the bunds, which will break up the bund and cause leakage, is not recommended. Plants such as banana, palm and papaya are acceptable and palms form wind breaks. The sides of the ponds must never be vertical. In any soil conditions this is bound to lead to the rapid breakdown of the bund. Figure 17 illustrates the result of this practice.

The tops of the bunds between ponds should be about 1 m in width to allow workers to walk round the ponds carrying feed and harvesting gear. Figure 18 shows the result of combining too narrow a bund top with almost sheer sides: the bund has had to be staked to prevent collapse. This will need constant maintenance particularly as the site illustrated was sloping and the water level in adjacent ponds was quite different. The bund width must be increased to at least 2-3 m at one side of the pond (usually the drain end or where harvest nets are to be beached) so that trucks can be brought to the pond side for delivering post-larvae and feed and picking up harvested prawns. On larger farms, particularly where mechanical broadcasting of feed is employed, a wide bund top must be provided on one of the long sides of the pond as well as at one end.

7.2 Water

7.2.1 Supply

The characteristics of the water supply required for freshwater prawn farming have been discussed in Manual Section 6.2.

1/ Assumes a productivity of 2 000 kg/ha/crop

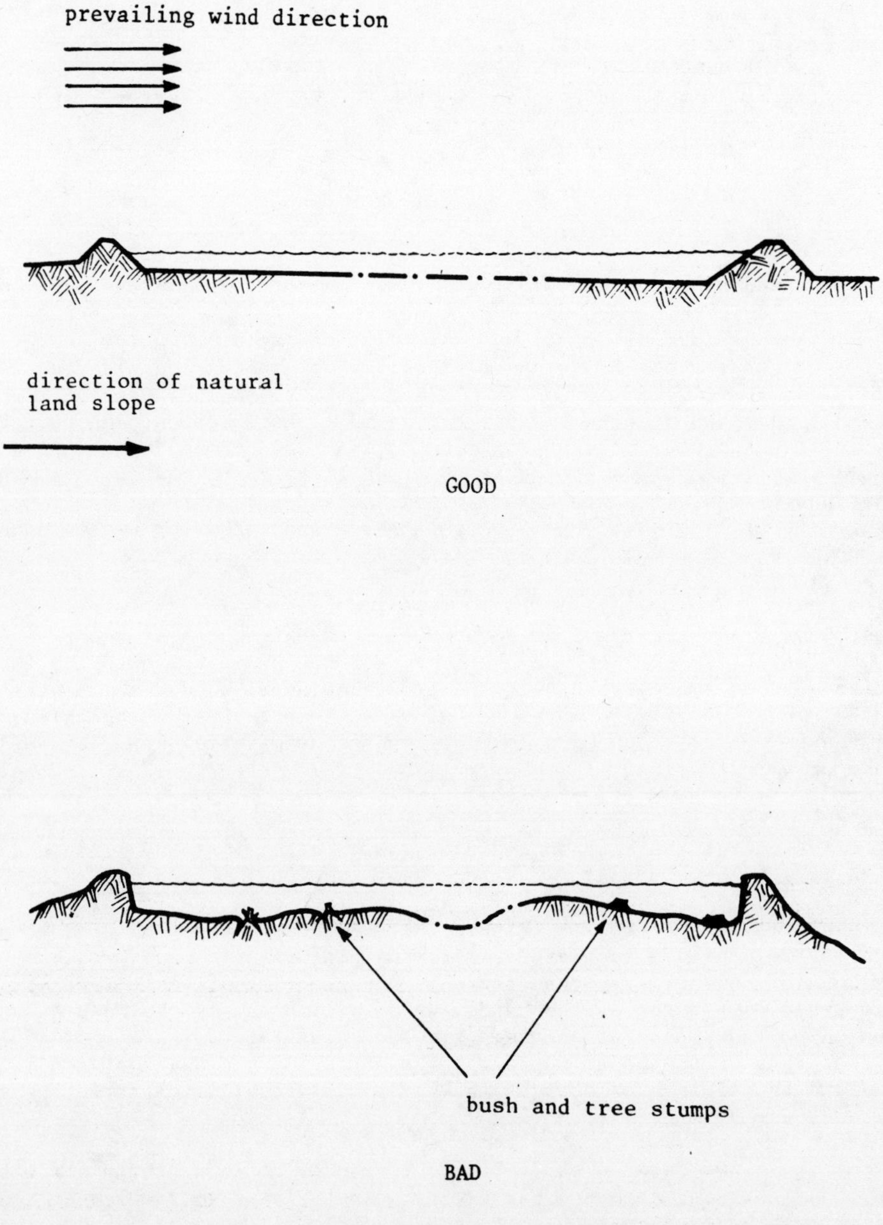

Figure 13. Pond bottom - good and bad profiles. Longitudinal section (not to scale)

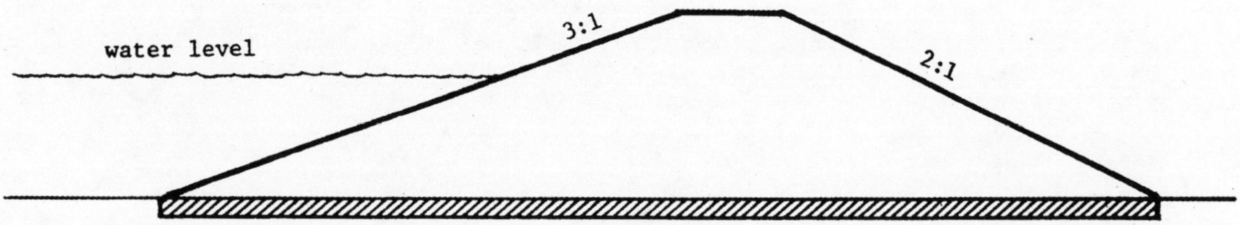

(a) Example of bund where bund constructed in an impervious area, with impervious material

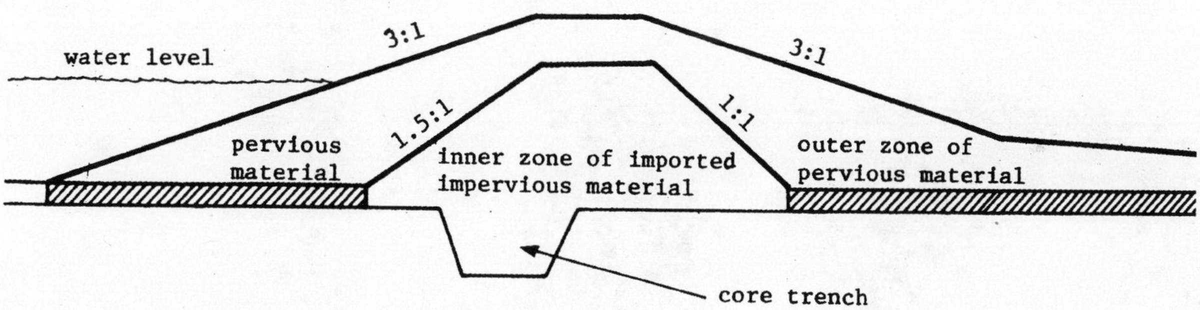

(b) Example of bund of pond constructed in a pervious area. Clay material is brought in from outside the site to form an impervious core to the bund which extends down to the impervious zone below the pond

The top soil and organic matter in this area should be removed before the bund is constructed

Figure 14. Cross sections of bunds (Sivalingam, 1974)

Figure 15. Bund erosion

Figure 16. Bund planting

This pond is within a few hundred metres of that illustrated in Figure 15. The owner has planted its bunds to help to prevent erosion. Fence against walking catfish is also shown.

Figure 17. Collapse of sheer sided bunds

Figure 18. Problems with sheer sided bunds with a narrow crest. Inner side of the pond had to be staked to prevent total collapse of bunds

7.2.2 Treatment

It is not normal to treat the water entering freshwater prawn ponds except to screen it to prevent entry of predators. Screening is not necessary where the water supply is piped from a well or a spring but is essential where surface water or open channel distribution is used. Well water requires aeration by cascading or by injection above pond water level to re-establish gas equilibrium, as it is often initially very low in dissolved oxygen content.

There are many alternative methods of screening incoming water, some of which are illustrated in Figure 19. Crude screening excludes adults and fingerlings of unwanted species but not their eggs or larvae. Figure 19 also shows a simple gravel filter which will exclude fish eggs and larvae as well.

7.2.3 Distribution

The way in which water is distributed and injected into freshwater prawn ponds is of great importance. Farms must be designed with a water distribution system that will allow the filling of one pond (or 10 percent of the pond surface area, whichever is the greater) at any time without starving the other ponds of replacement and flow-through water (Table 4).

There should not be any contact between incoming water and that drained from ponds. Each pond should have its own individual supply from a central water distribution channel and should not receive the outflow from another pond. The transfer of water from one pond to another is not recommended since it means poorer water quality conditions in the second (and subsequent) ponds and brings the risk of disease transfer.

Ideally water should be distributed in pipes or open channels by gravity if the topography of the site allows it (Figure 20). Similarly, inlet pipes or channels should be constructed above the water level in the ponds so that incoming water falls onto the surface of the water (Figure 21). This may be achieved by pumping the water supply to an elevated channel if this is economically feasible. The water inlet is normally placed at the shallow end of the pond, diametrically opposite to the discharge point. The inlet channels/pipes must be correctly sized according to the water demand of each pond. Further information on this topic can be obtained in Sivalingham (1974) and Wheaton (1977). Table 6 gives average discharge rates for pipes of different diameters. The water requirements for various sizes of ponds are given in Table 4.

The flow of water into each pond must be controlled by valves, weirs, stop-logs or plugs (Figure 23).

While gravity supply, elevated water inlets and lack of cross-contamination of water between adjacent ponds represent the ideal, many freshwater prawn farms exist which do not comply with these recommendations. Many farms, due to site, technical, or financial limitations, have water inlets below the pond water level (Figure 22) and receive water from an inlet channel with the same water level as the pond. In some cases ponds are directly interconnected. These farms produce freshwater prawns, often in substantial quantity (New, 1982) and, at least in Thailand, profitably (Shang *et al.*, 1980). The authors of this manual simply wish to point out the dangers of such practices and to suggest that higher production levels are more feasible if it is possible to have an improved water distribution system.

7.2.4 Discharge

It is preferable to be able to drain ponds by gravity than to have to pump the water out and, where this is possible, a monk, or sluice gate outlet structure should be constructed. These structures (Figure 24) allow the farm operator to easily control water depth and drainage speed and are screened to prevent the loss of stock. In flow-through water management, water is continually flowing through this structure.

If the "continuous" rather than the "batch" system of culture is practised it may rarely be necessary to totally drain the pond but the presence of a "monk" or sluice gate provides this facility if required. More important, it enables water level control during seine harvesting operations, flushing and water circulation.

(a) Pipe screen

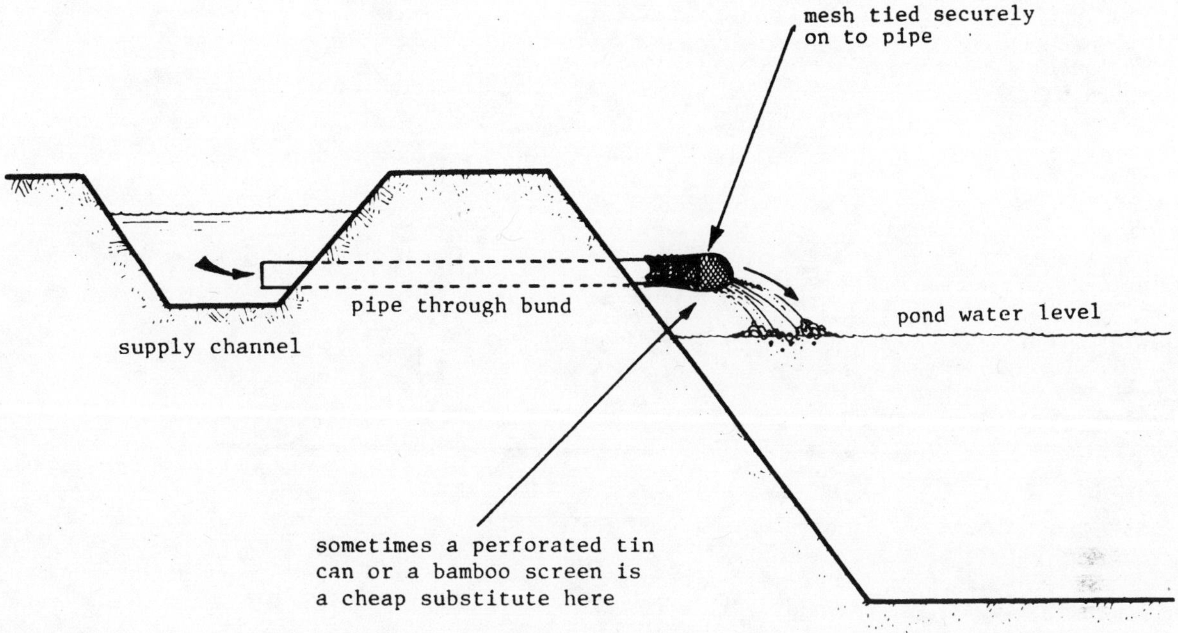

(b) Filter box

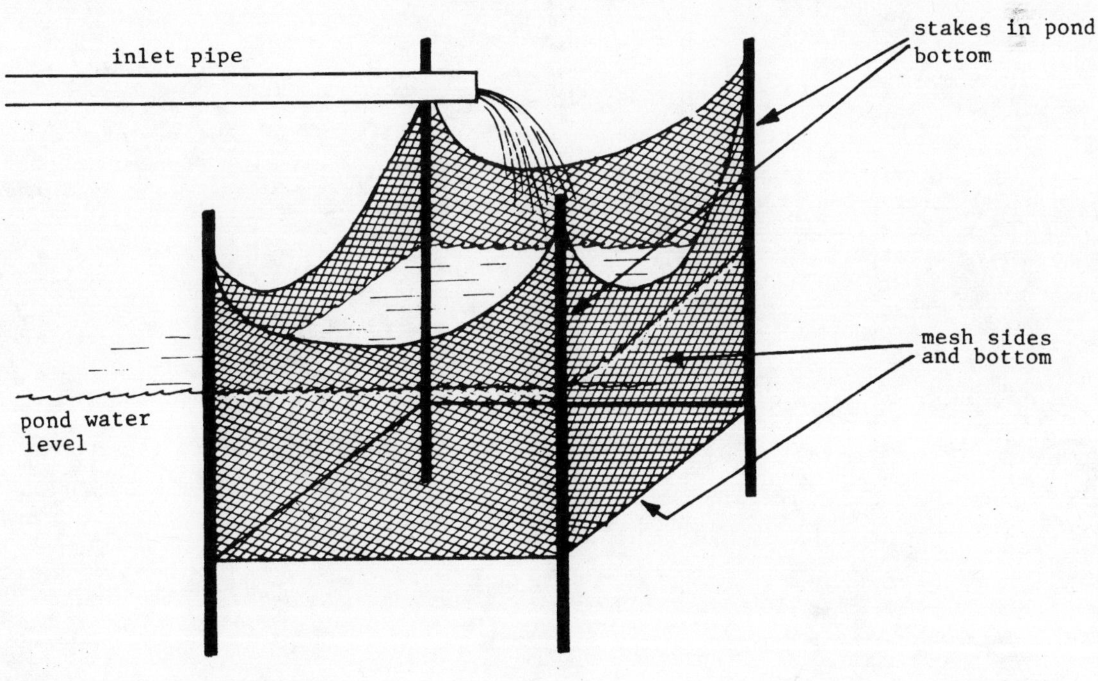

Figure 19. Inlet water screens

(c) Horizontal screen (placed in water inlet channel)

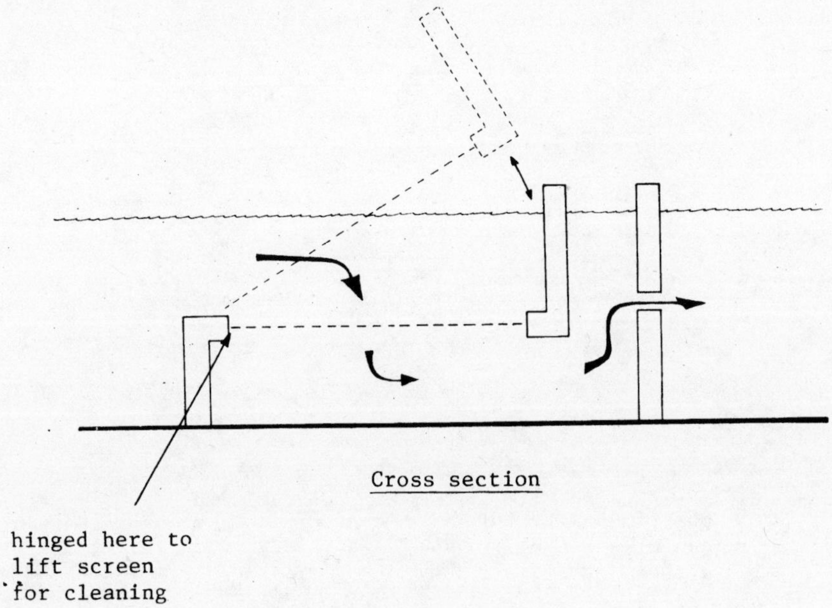

(d) Vertical screen box or channel

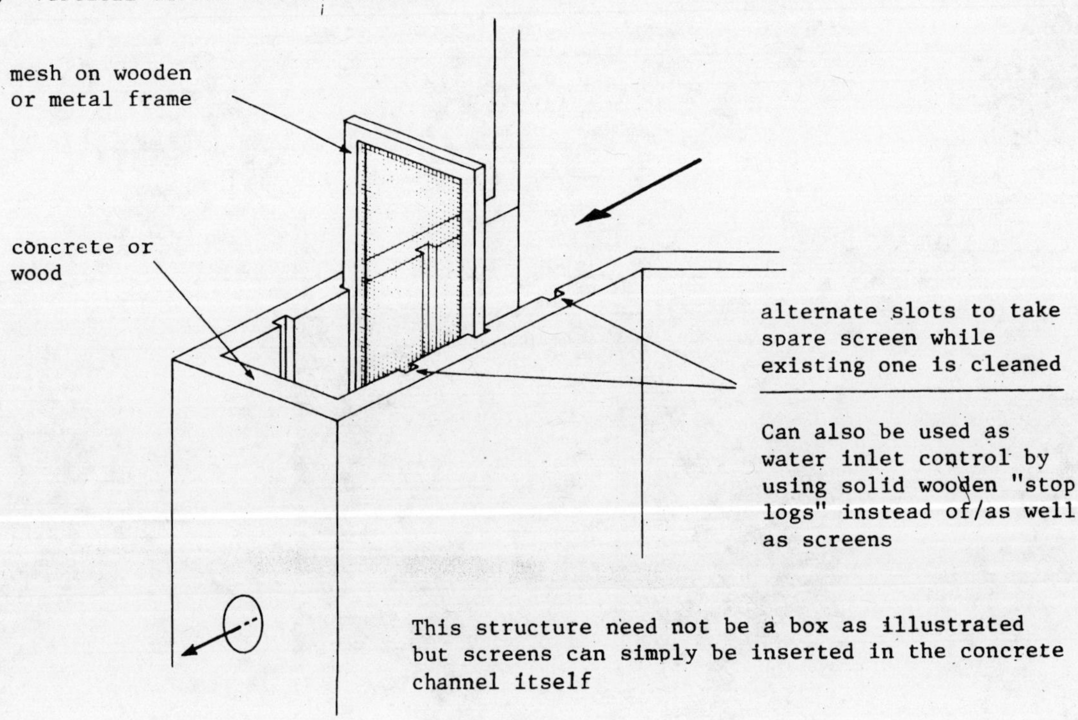

Figure 19 (continued).

(e) Gravel filter (after Sivalingam, 1974)

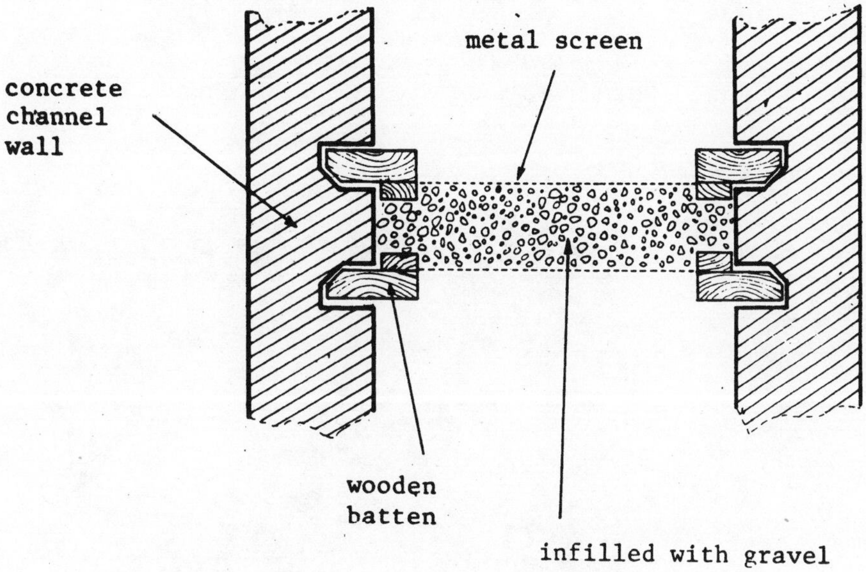

Viewed from above

If two positions for these are built in each inlet channel, a new filter can be inserted before the old one is taken apart for cleaning

Figure 19 (continued).

Figure 20. Gravity water distribution

Figure 21. Water inlets above water surface level

Figure 22. Water inlets below water surface level (pond under construction when photograph taken)

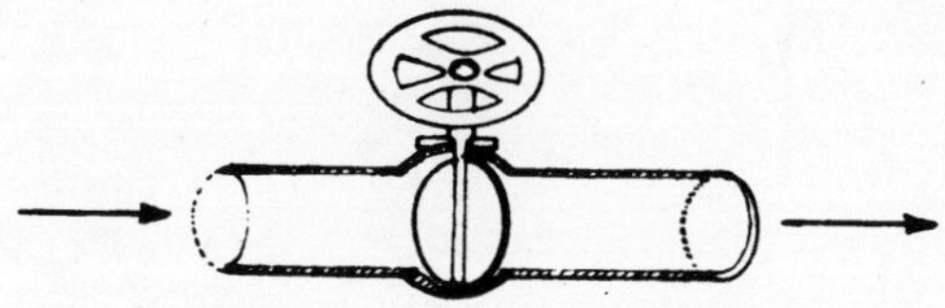

(a) Pipe and valve

(b) Sluice (stop log)

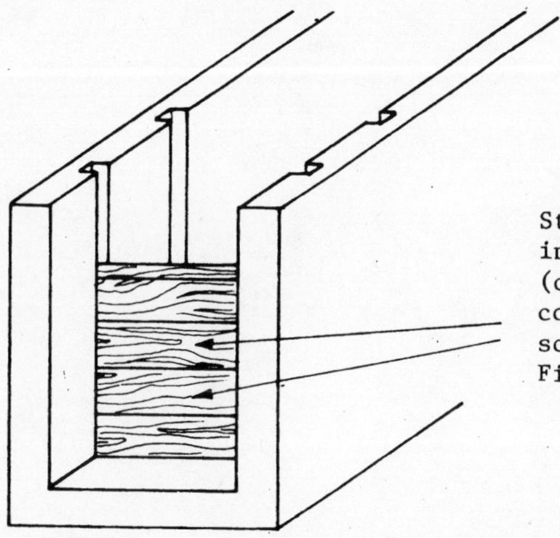

Stop logs in grooves in concrete channel (can be used in conjunction with screens - see Figure 19)

(c) Pipe with plug (also plugged by tying impervious bag or sheet over the pipe)

Figure 23. Water inlet control

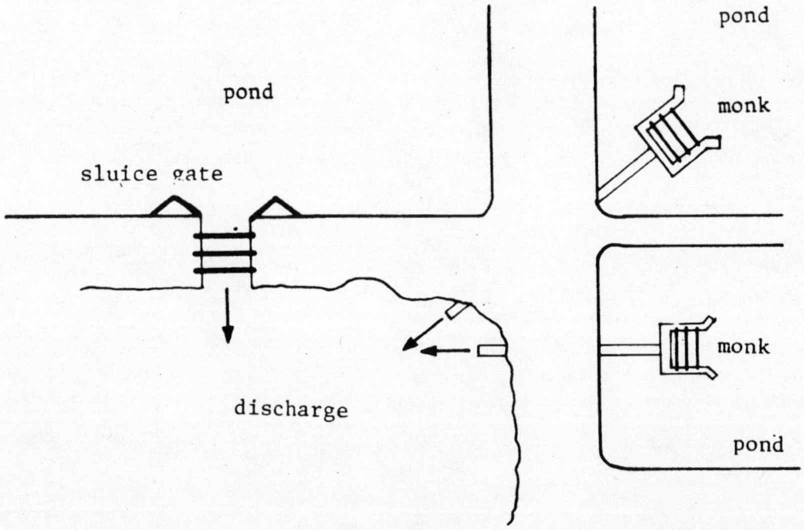

The sluice gate is in the bund — The "monk" is inside the pond

Both have similar principles:
a "monk" is illustrated below

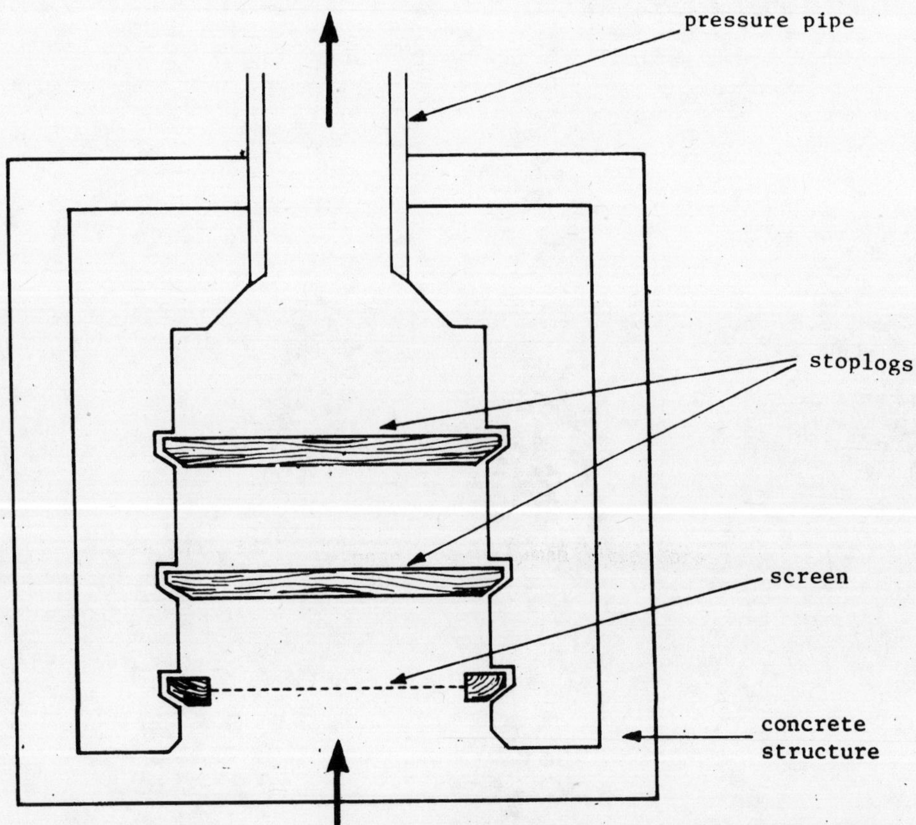

view from above (after Sivalingam, 1974)

Stop logs and screen are interchangeable

Some monks (or sluices) have 4 sets of grooves for versatility (to enable easy screen changes)

Figure 24. Sluice gate and "monk" design. No dimensions given; only principle is illustrated

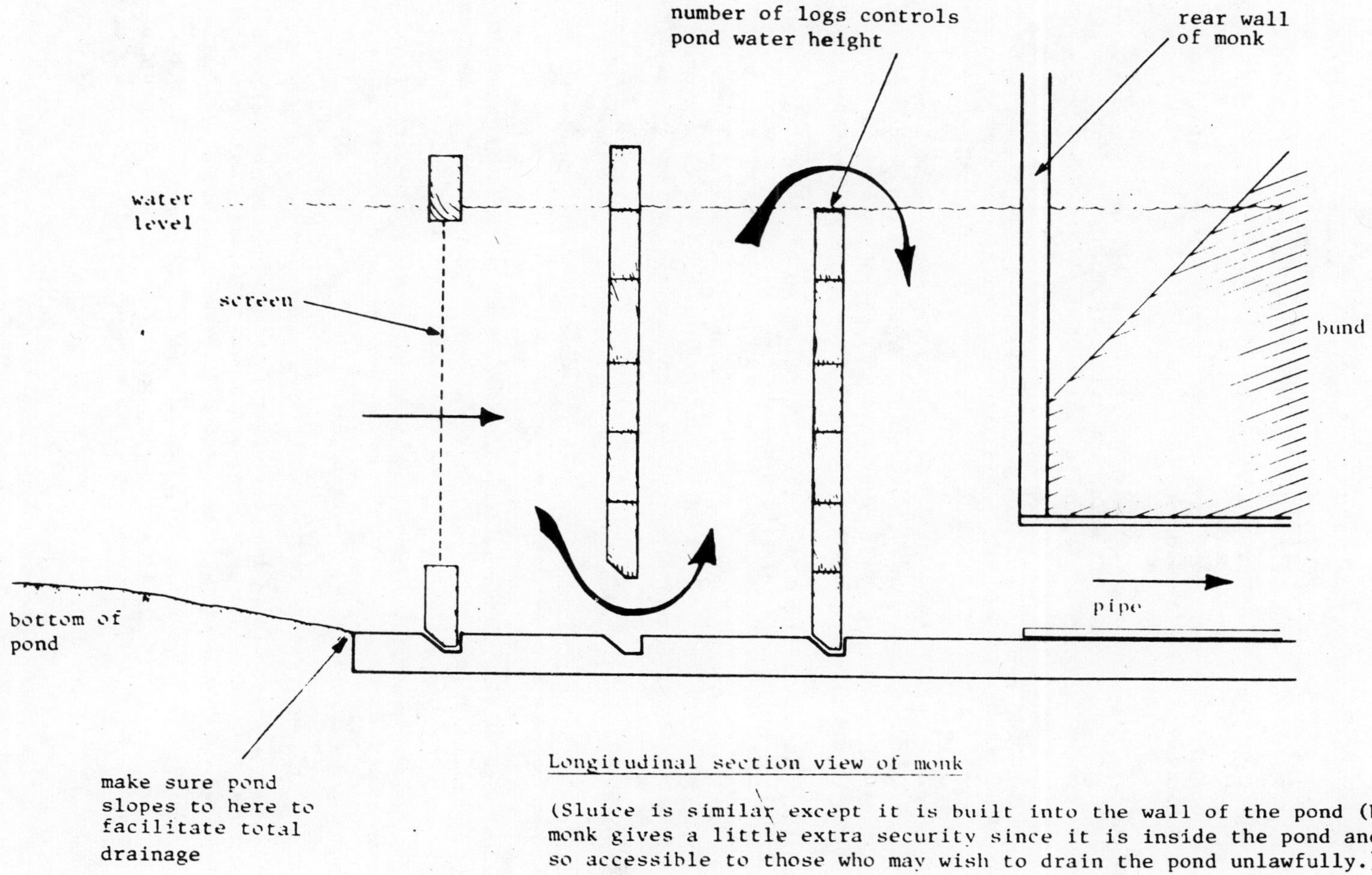

Longitudinal section view of monk

(Sluice is similar except it is built into the wall of the pond (bund); monk gives a little extra security since it is inside the pond and not so accessible to those who may wish to drain the pond unlawfully.)

Two sets of stop logs enable water to be drawn from the bottom of the pond, as shown above; one set of logs is okay but only allows flow from the pond surface.

Figure 24 (continued).

Static (non-flow-through) ponds can have a simple screened and plugged outlet pipe or a turn down drain similar to that used in larval tanks (Figure 2). Outlet structures, whether they be pipes or "monks" must be carefully sized so that the pond does not drain too slowly (Table 6), and they should be sited so that the pond can be totally drained (Figure 24). Figure 25 is an illustration of a sluice gate structure. The top of the sluice gate should also be constructed at least 50 cm above the highest pond water level as a safety measure.

Where the pond outlet is a pipe below water level, there should also be an overflow pipe inserted about 20-30 cm below the top of the bund, above the normal water level in the pond. This overflow pipe should be screened in the same way as the normal pond outlet, to prevent loss of stock. If the water level in the area to which the pond drains also rises, however, the overflow pipe will be ineffective.

Where drainage by gravity is not feasible because of the limitations of the site, the only way to empty the pond or control its water level is by pumping. A screened "long-tail" pump is one method of emptying ponds on flat sites which is used in Thailand (Figures 26,39,41). These pumps are readily available since they are used for paddy-field irrigation.

The topics of harvesting and harvest structures are dealt with in Manual Section 8.4.

7.3 Aeration

Permanent aeration is not normally practised in freshwater prawn ponds. However, the dissolved oxygen level of incoming water is increased if ripples (Figure 27) are built into gravity inflow channels and water is injected into the ponds above water level (Figure 21)

Most prawn farms use water exchange as the favoured technique for curing low dissolved oxygen levels, as well as other water quality problems.

A number of other techniques, including recirculation by pumping, and aeration through the use of floating (Figure 28) and submerged (Figure 29), and paddle wheel (Figure 30) aerators, are also available for emergency aeration when dissolved oxygen levels are low.

7.4 Miscellaneous

In addition to its basic ponds and water distribution systems a freshwater prawn farm has the following equipment and facility requirements:

(a) storage: dry storage facilities for feeds (or raw materials to be used for feed manufacture), chemicals, nets, etc. A 40-ha farm taking monthly deliveries of feed, for example, will require dry and cool storage for up to 20 t of feed;

(b) accommodation: every farm should have accommodation for some of its workers to live on site;

(c) nets: see Manual Section 8.5 and Appendix 8;

(d) water quality equipment: see Manual Section 4.6;

(e) fencing: a perimeter fence and, on larger farms, lighting, is advisable to deter poachers or human "predators". Placing 60 cm high netting, with part of the netting buried below the soil surface, around the pond area prevents invasion by adult catfish and snakeheads, in areas where they exist, (Figures 11, 16);

(f) transport: larger farms will need their own trucks for the collection of post-larvae for stocking, the delivery of marketable prawns and the collection of feed.

8. FARM OPERATION

8.1 Pond Management

As mentioned before, vegetation along the pond bank or "bund" minimizes bank erosion. Below the water line, it also provides food and a habitat for the prawns. Care must be taken

Figure 25. Sluice gate

Figure 26. Longtail pump

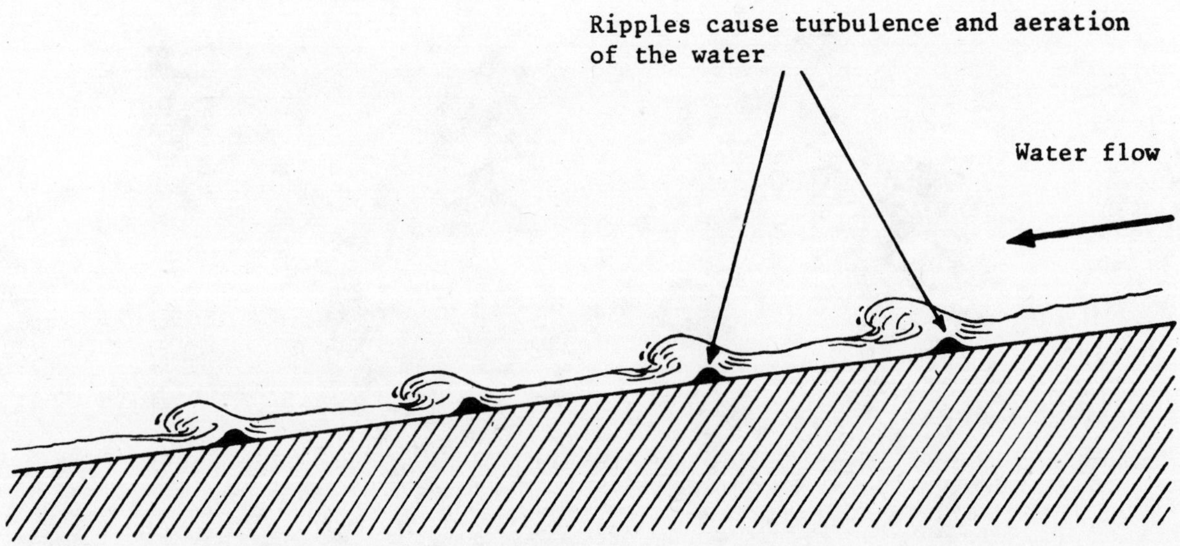

Figure 27. Water ripples

Figure 28. Floating aerator

Figure 29. Submerged aerator

Figure 30. Paddle wheel aerator

however, that the growth does not become so excessive as to interfere with harvesting. The pond depth must be maintained at an average of 0.9 m and there should be no extensive shallow areas, or aquatic-rooted plants will grow extensively on the pond bottom which is undesirable from a harvesting point of view (Figure 31). The growth of aquatic-rooted plants, and benthic algae must also be discouraged by management practices (Manual Section 8.3) which encourage significant growth of phytoplankton, thus reducing light penetration to the pond bottom.

When a pond which has previously been stocked with fish is to be converted to freshwater prawn culture or, when fish are known to be present, the pond should be treated with a piscicide. Rotenone, used at 1 to 2 g/m³ is effective if spread evenly throughout the pond 1/. Other more powerful chemicals are sometimes used because of their cheapness but their use cannot be recommended because of the danger to humans. Crude saponin, water extracted from tea seed (*Camellia* sp.) has been effectively used at 1.1 ppm in experiments in mixed populations of marine shrimp and fish. This level removes predatory fish from saline water without harming the crustaceans present (Terazaki *et al.*, 1980), but its use in freshwater prawn culture is not yet reported.

New ponds should be limed. The quantity of lime used depends on the soil characteristics and soil analysis is necessary for accuracy. This topic, and that of fertilization, is a complex one and is not specific to freshwater prawn culture. The reader is suggested to seek the advice of his local fishery officer and to consult the text by Boyd (1979). In practice in Thailand, a standard application of 1 000 kg/ha of agricultural limestone is recommended each time a freshwater prawn pond is drained.

Where the pond is built on acid sulphate soils corrective measures using lime are generally not recommendable, due to the high lime requirements of these soils. Liming should be limited to the banks of the pond and should be combined with planting of acid resistant grasses such as the African star grass. Continuous flushing of the pond water and over the banks of the ponds, followed by drying, accelerates the reclamation process of this type of pond. The period required to correct pH may vary between a few months and several years, depending on soil and climate characteristics. It is not recommended to build ponds on suspected acid sulphate soils as their recovery is expensive and laborious. Generally the productivity of ponds improves as they get older and as a rich bottom area and grassy banks are established. Ponds having a high water pH can be improved by "ageing". This means filling with water 2-4 weeks before stocking and allowing natural biological processes to buffer the pH.

After liming, the ponds are filled prior to stocking. Fertilization is rarely necessary in freshwater prawn culture as an adequate phytoplankton density, providing cover and controlling weed growth is quickly provided by the feeding regime (Manual Section 8.3). However, ponds built in a sandy-clay soil may require fertilization as such. Where necessary, 25 kg/ha/month of triple superphosphate will keep the water green.

The operator should take care to maintain his ponds well during the farming period. Special attention must be paid to the prevention and treatment of bund erosion, the control of aquatic-rooted plants, and the maintenance of water inlet and outlet structures, particularly the screens. The plants *Elodea* spp. and *Hydrilla* spp. make a good substrate for prawns.

The topic of water management is dealt with in Manual Sections 8.3 and 8.7.2. Some farmers using static water rather than continuously flowing water, make a practice of changing water regularly every two weeks in an effort to promote synchronous moulting of post-larvae.

The surface area of a pond, available to the prawns, can be increased by placing rows of netting, suspended from floaters and weighed down with sinkers, across the pond. The use of twigs, pipes, bricks, etc., as prawn habitats interferes with the harvesting process.

1/ Commercial preparations of rotenone containing 5 percent of active product

Experience shows that it may be necessary to "re-condition" ponds which have been in continuous operation for several cycles. "Re-conditioning" is carried out by complete draining, followed by sun drying and filling to aerate the pond bottom. Care must be taken during this operation to minimize possible damage to the water retention characteristics of the pond.

8.2 Stocking

Post-larvae may be stocked into the ponds as soon as they are filled with water. Usually post-larvae which are only 1-4 weeks old (after metamorphosis) are used to stock the ponds, where they will remain until harvesting. An alternative procedure is dealt with in Manual Section 8.8.1. Some farmers prefer to use post-larvae reared in a simple rather than a sophisticated hatchery, believing them to be more hardy.

On arrival at the pond bank care should be taken to acclimatize the post-larvae to the temperature of the pond by floating the transport bags in the pond for 15 minutes (Figure 32) before emptying them into the water (Figure 33). There is evidence (Sarver *et al.*, 1982) that a significant difference between the pH of the water in which the post-larvae have been stored and transported and that of the pond can cause mortality as severe as thermal shock. The pH of the water in the pond to be stocked should be checked before the post-larvae are shipped so that they can be gradually acclimatized to the new pH over a 1-day period in the hatchery holding tanks.

The stocking rate used depends on the market size desired and on the management of the pond, particularly the harvesting procedure. In Thailand, where the favoured size for marketable prawns is about 70 g (head-on), and many farms have a growing season limited to about 8 months due to seasonal water supply, we recommend a stocking rate of 5 post-larvae/m^2 (50 000/ha). The post-larvae used are up to one month old (following metamorphosis). Some farms use a higher stocking density, even as high as 20/m^2 (200 000/ha), normally with the result that their production is higher (but not proportionately so), and that average prawn size at harvest is reduced. Where the "continuous" method of culture (Manual Section 8.5) is used, higher stocking rates are recommended, between 16 and 22/m^2 (160 000 and 220 000/ha) per year. An example of a stocking strategy for this technique is given in Appendix 9.

The post-larvae will have been counted into the transport bags at the hatchery; often the farmer is present at this time to ensure fairness. Normally hatcheries will overship post-larvae rather than underestimate. If the farmer is receiving post-larvae without having seen them packed, it is advisable for him to count the contents of one bag to check the accuracy of the delivery. Usually post-larvae are packed 1 000 or 2 000 to each bag so it is only necessary to count the number of bags allotted to each pond to stock at the desired density. Although the efficiency of stocking mosquito fish and guppies in the ponds to control predation by dragon fly nymphs on young postlarvae is doubtful, some extensionists recommend it as useful.

8.3 Feeding

8.3.1 Feed type

While some production, perhaps 200-300 kg/ha/year, may be achieved solely by relying on the natural productivity of the ponds, successful commercial farming of freshwater prawns must involve supplementary feeding. The types of feed used vary widely including individual animal or vegetable raw materials, feed mixtures prepared at the pond bank (Figure 34) and compounded feeds, bought from feedstuff manufacturers.

In Thailand, rice and rice byproducts form a significant feed ingredient of mixtures made by farmers for freshwater prawns. Cassava or tapioca is another major vegetable material used, while "trash" fish, molluscs, and prawn waste form valuable animal protein sources Meal made from the leaves of the Ipil ipil bush *Leucaena* sp., has formed a constituent of shrimp and prawn diets both in Thailand and Tahiti though its use is cautioned by the toxicity of mimosine, which is a problem in its use for terrestrial animals. Some farmers add individual raw materials to their ponds as well as, or instead of, mixed feeds. These include pig manure and the mortalities from chicken farms which, staked out around the periphery of the pond (Figure 35) seem to provide a delicacy for prawns.

Figure 31. An example of severe 'weed' (in this case, rice) growth in a pond which is too shallow

Figure 32. Acclimatizing postlarvae in transport bags before stocking

Figure 33. Stocking postlarvae

Figure 34. Feed preparation at the pond bank

Figure 35. The use of chicken carcasses as fresh-water prawn feed

Generally speaking, the use of individual raw materials, especially wet materials such as trash fish, pose more of a potential pollution hazard than mixed or compounded feeds (especially where the latter are water stable). Compounded chicken feeds, either unmodified, or re-extruded through a mincer with trash fish or prawn meal, have been used with success in freshwater prawn farming.

Knowledge of the specific nutritional requirements of shrimps and prawns is now increasing (New, 1976, 1980a, 1980b), and some compounded feeds designed mainly for marine shrimp, which are more water stable than chicken feeds, are being used in freshwater prawn farming. The use of feeds designed for other animals such as pigs or chicken, for feeding prawns, has some dangers because of the unknown effects of some of the growth promoters and drugs used in these rations.

Diet water stability, which enables the prawn to receive a balanced ration, rather than to select individual ingredients also lessens water pollution and provides a ready method for the farm operator to judge his daily pond feeding rate. Water stability can be imparted to feeds by a wide range of naturally occuring and modified gums and binders, by the inclusion of pre-gelatinized starch and by certain processing techniques used by feed manufacturers.

Compounded diets designed primarily for marine shrimp which seem suitable for freshwater prawns are manufactured by several companies including Ralston Purina 1/ and Central Soya 1/ made in the USA, but available in S.E. Asia and Charoen Pokphand and Sahapatanakaset 1/ in Thailand. Most Japanese and Taiwanese-made shrimp feeds are designed for marine species with a much higher (around 40 percent) protein content than is required for freshwater prawns and are therefore generally too expensive for this purpose.

Compounded diets may seem expensive (at the time of writing, in 1981 they vary, depending on location and formulation, from about US$ 400 to well over US$ 1 000 per t). However, they are generally more efficient to use than other feeds and, when the basic success criterion (food conversion ratio) is applied, often prove cheaper. When considering the choice of feed what really matters is:

(a) what is the total productivity of the pond achieved through its use and,

(b) what are the total feeding costs (including the feed itself, its transport, feeding, and the problems it may cause in pond management) of producing each kilogramme of marketable prawn.

Food conversion ratios (weight of feed presented divided by weight of animals produced) of 2:1 to 3:1 may be expected for compounded diets. The food conversion ratio of wet materials, such as trash fish, will be much higher because of their moisture content - perhaps 7:1 to 9:1.

Though the use of compounded diets may be ideal, they are not always available to the small farmer, who may also have local access to acceptable and cheaper feed ingredients which he can prepare himself. The farmer may also have problems in storing compound feedstuffs in humid conditions where deliveries cannot be made regularly in small quantities. Some typical formulae for freshwater prawn diets are given in Appendix 7, which also discusses feed preparation and gives some information about commercial shrimp feeds.

8.3.2 Feeding rate

There can be no exact general recommendation for daily feeding rates, since these depend on the size and number of prawns (and fish) in the pond, the water quality, and the nature of the feed. Good practice is for the farm operator to feed to demand. Feed is normally spread around the periphery of the pond in the shallows which are good feeding areas. Sometimes the feed is presented in defined "feeding areas" a few metres apart. Both practices enables the farm operator to see how much feed has been consumed.

If there is no feed left on the following day, the feeding rate should be increased. If there is excessive food left, the feeding rate should be decreased or feed even omitted

---

1/ The inclusion of a specific brand in this manual neither implies recommendations nor the absence of alternatives

for one day. This need for the operator to be able to see unused feed after 24 hours highlights one of the assets of a water stable diet.

We recommend that the initial feeding rate, using a dry diet such as a compounded chicken feed in a pond stocked at $5/m^2$, should be about 6.25 kg/ha/day. This is far more than the prawns will consume when they are young post-larvae but the diet also acts as feed for the mosquito fish. It additionally acts as a fertilizer to build up the plankton density to a level which will provide cover for the prawns and prevent the growth of aquatic-rooted plants. The use of feedstuffs to induce phytoplankton growth is an expensive but simple and effective technique. Later, when the effect of this process has been demonstrated, the use of fertilizers to replace most of the feed in the first two months could be encouraged. The use of nursery ponds (Manual Section 8.8.1), where the prawn density is higher, also results in a more economic use of feedstuffs.

Over-feeding is continued until the phytoplankton density becomes such that a Secchi disk reading shows (Figure 6) a visibility of between 25 and 40 cm. A cruder method of making this measurement is to put your arm in the water. If you can easily see the tips of your fingers the water is too clear. If you cannot see your wrist then the phytoplankton density is too high.

Where flow-through is possible all the time, the phytoplankton density can be controlled by altering the water flow rate. Even with static ponds it is possible to flush a pond out if the phytoplankton density becomes too high (or there are other reasons to suspect water quality) by partially draining and refilling the pond, if water is available. Normally the best means of controlling phytoplankton bloom is by carefully monitoring the effect of feeding rate and altering it as necessary. By this means panic situations caused by gross over-feeding can be avoided.

From the time that the phytoplankton density reaches the desired level the daily feeding rate is determined by the daily consumption of the prawns (demand feeding), together with an examination of the colour and transparency of the water, as noted above. The pond bottom can be inspected by means of a plastic tube with a mirror like a reversed periscope. Once-daily feeding in the late afternoon is recommended although many farmers prefer to split the daily ration into two feedings, early morning and late afternoon.

The daily feeding rate will rise gradually from the initial (6.25 kg/ha/day) to a much higher level at harvest time. The exact level of feeding depends on the growth and survival rate of the prawns and the standing crop in the pond at any one time. As an example, you can expect the feeding rate to build up to a level as high as 37.5 kg/ha/day at the time just before a pond production of 1 250 kg/ha is batch harvested (after 6-8 months).

Similarly the feed consumption of a pond which is being "continuously" operated (regularly cull harvested: Manual Section 8.5), may at times reach as high as 45 kg/ha/day if producing 3 000 kg/ha/year. The average daily consumption of a pond producing 2 500 kg/ha/year, for example, should be 14-21 kg/ha/day.

8.4 Monitoring

Methods of monitoring feeding rate and phytoplankton density and for the control of the latter have been dealt with in Manual Section 8.3.2. It is good practice, if possible, also to monitor other water quality parameters such as pH, temperature and dissolved oxygen (Manual Section 4.6) so that production rates can be linked with the environment of each pond and so that remedial action (as in the case of low dissolved oxygen levels, for example) can be taken. Juvenile freshwater prawns will tolerate early morning dissolved oxygen levels of 1-1.5 ppm (Wulff, 1982), but it is not advisable to allow extended periods at such levels.

It cannot be over-stressed that the farmer should keep adequate written records of such things as water quality, stocking rate and date, daily feeding quantities, dates on which water changes are made (and how much), harvesting dates and quantities, etc. Only in this way can he build up a picture of how each pond behaves under a certain management regime (and every pond is different) and accurately apply his experience to future pond management to operate his farm profitably. This applies equally to hatchery management.

Ideally, a farmer should be able to determine the average size and the number of prawns in his pond at any time. In this way he could tell whether growth and survival rates were satisfactory, or not, and determine a daily feeding rate based on a percentage of the pond biomass. Even if this were possible, daily feeding rates based on a percentage of biomass should not be applied blindly but should be tempered by the observations on consumption and phytoplankton density referred to in Manual Section 8.3.2. Unfortunately (Appendix 6) there is no accurate way known of determining the standing crop of freshwater prawns in a pond unless the pond is regularly seined (as is done in "continuous" operation).

One serious problem encountered in extension work with farmers that have not grown freshwater prawns before is that their enthusiasm declines rapidly after stocking time. This is mainly because they cannot see the prawns after they are stocked; they are difficult to see and to catch at this stage. At this time the pond is being "over-fed" with quite large quantities of food and, after a while, the farmer begins to think that the feed (and his money) is being wasted because all the prawns must have died, escaped, or been eaten by predators. He then decreases or (worse still) stops feeding until, about two months after stocking, he begins to see (by now quite large) prawns again. He then starts feeding again but by then the productivity of that crop has been permanently reduced. This is a common experience and the farmer needs a great deal of encouragement at this time. He should also be taught to examine the perimeter of his ponds by night, with the aid of a flashlight.

Those who are about to grow freshwater prawns for the first time must be made aware of the fact that there is a wide divergence in growth rate within a population of prawns. Some will grow very fast, others hardly at all. This is a normal characteristic of the animal. This disparity in growth rate is more pronounced among males than females and in mature populations of freshwater prawns in ponds, three major size classes exist. These are large "bull" males, females, and small "bachelor" males. If length-frequency graphs are constructed which differentiate between males and females it will be seen that those groups are demonstrated by three peaks (Figure 46). A method for sexing small (juvenile) prawns is shown in Figure 1g (Pedini, M., personal communication, 1981).

Males weigh slightly more than females of the same length, but not markedly so. Growth rate may be measured either by weight or total length. Though measurement of length from the eye orbit to the tip of the telson is more reliable (because the rostrum of some animals becomes shortened by damage) in farming practice total length, from the tip of the rostrum to the tip of the telson, is usually measured, often by ruler (Manual Section 8.5). Figure 47 gives the relationship between total length and live weight for a mixed-sex population of freshwater prawns.

The growth rate and survival of a population of prawns depends on many factors, including density, predation, feed and temperature. Since these factors are so site and operator specific it is not wise to predict what growth rates should be, for fear that these will be too rigidly applied. However, the following examples are provided from Willis and Berrigan (1977). In earthen ponds with temperatures ranging from 20.5 to 30.5°C and with a mean of 27°C, juvenile prawns stocked at a mean weight of 0.78 g (orbit to telson mean length 31.7 mm) had a mean weight at (batch) harvest of 43.26 g (orbit to telson mean length 108.0 mm) after 167 days in the ponds. In another three ponds, where post-larvae instead of juveniles were used, prawns stocked at a mean weight of 0.055 g (mean length 14.5 mm) grew to a mean weight of 28.2 g (mean length 95.2 mm) when harvested after 170 days. The survival rates were very high at 79 percent for the juveniles and 88 percent for the post-larvae. In farming practice survival rates of 50 percent between metamorphosis and (batch) harvest could be acceptable.

8.5 Harvesting

The time of harvesting depends partly on growth rate and the market size desired and partly on the pond management technique chosen. Basically there are two extreme methods of managing a freshwater prawn pond. One, the "batch culture" technique, is to stock the pond, allow the animals to grow until the average market size is achieved or the pond has to be drained for other reasons (for example, shortage of water; low temperatures), and then harvest the whole crop. The other technique "continuous culture", consists of stocking the ponds, usually once a year, at much higher densities than with "batch culture" and usually after 5-7 months, depending on growth rate (temperature) and the local marketable size, culling off market sized animals by seining at regular intervals.

In this "continuous culture" technique, the ponds are never drained and market sized animals are removed by seining. The ponds are normally totally seined once a month (or half of the pond is seined twice a month) to avoid disturbing the whole pond too frequently. Some farms restock with post-larvae once per year; others on a more frequent basis (four or six times per year). This "continuous culture" or "continuous culling" technique is favoured by many farmers in Hawaii and is being tried by some of the larger farms in Thailand. Its protagonists claim that it is the method which maximizes production, achieving outputs of 2 500-4 000 kg/ha/year. An example of management strategy for this technique is given in App.9.

Some large farms use the "batch culture" technique effectively, and one farm in Honduras (Wulff, 1982) reports that both techniques show promise but that neither has a clear advantage over the other. This farm averages a production rate of 3 000 kg/ha/year of freshwater prawns but their marketable size is only 30 g (head-on) and the animals are marketed as tails only.

In Thailand, most farmers use a combination of the two techniques. About five months after the post-larvae are stocked, cull harvesting commences. The pond is totally culled once per month or partially twice per month. Market-sized animals are taken out and sold while smaller ones are returned to (or remain in) the ponds for further growth. After about 8 months the pond is drained and the whole harvest sold. The pond is then treated, refilled and restocked immediately, or remains empty until the water supply becomes available again.

In all cases harvesting operations should take place as early as possible in the morning when it is cooler, to avoid having pond levels too low when the sun is directly over head. The prawns will be subjected to extreme rises in temperature and decreases in dissolved oxygen in shallow water and there will be many mortalities before the animals can be harvested.

Harvested prawns can be divided into a number of groups:

(a) large or "bull" males (about 50 percent of the males);

(b) small males which do not contribute to the cull harvest in the first year of continuous operation;

(c) females, either egg-bearing or not;

(d) soft shelled (newly moulted) prawns;

(e) "terminal growth" prawns.

Good quality harvested prawns are greenish with bright blue or yellow chelipeds (claws).

8.5.1 Cull harvesting

In this technique a seine net is pulled through the pond to remove harvest size animals from a "continuously operated" pond. The net may be a simple seine or one constructed especially for the purpose (Appendix 8). The mesh chosen for the net will depend on the size of animal to be marketed but 1 1/2 in (3.8 cm) to 2 in (5 cm), stretched, is normal.

Usually the first cull harvest in a new pond to be "continuously operated" takes place 5-7 months after initial stocking (but this depends on the market size being sought and the growth rate, which is mainly influenced by temperature and feeding, in each pond).

Care must be taken to ensure that the bottom of the seine is kept on the pond bottom, or many prawns will escape beneath it. Preferably the seine should be pulled down the long axis of the pond (this is why rectangular ponds of 30-50 m maximum width are preferred: Manual Section 7.1.1) so that the ends of the net are pulled along the banks of the pond (Figure 36a). Seining different halves of the pond once every two weeks avoids disturbing the whole of the pond at once.

The amount of prawns collected by the seine should not exceed that quantity which can be rapidly taken out of the seine, and transferred to a live box or a cage or impounded area for sorting. Many prawns, especially smaller ones, die when the bag of a bag seine is lifted out of the water if the quantity is too great. One way, practised on a farm in Honduras (Wulff, 1982) to ensure that the seine does not get overloaded when the whole pond is being

(a) Single seine operation

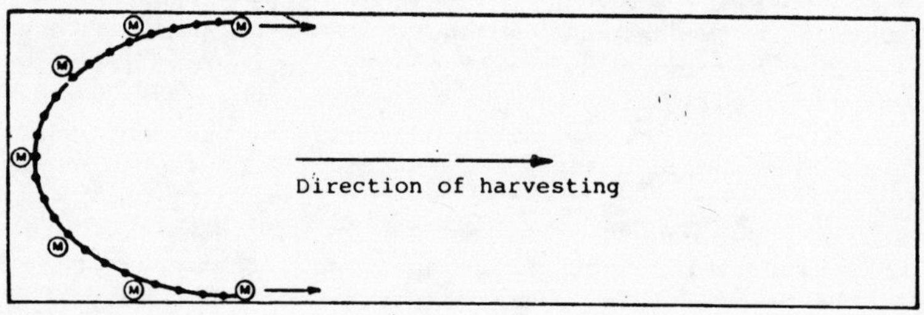

Ⓜ = man

(b) Double seine operation

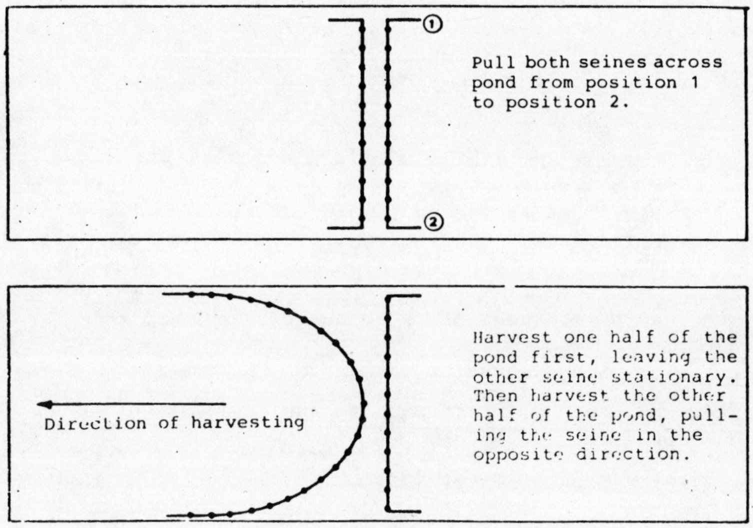

(c) Seining a wide pond

M = this man beats the surface of the water with a stick and moves in the same direction as the seine is being pulled, a little in front of it but further out into the pond.

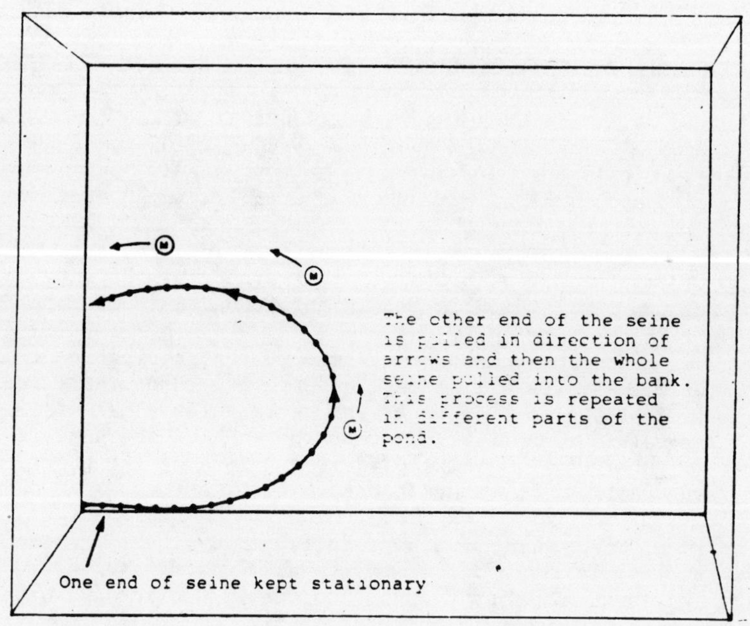

Figure 36. Pond seining

seined is to draw two seines across the middle of the pond. One seine is drawn to one end and harvested. Then the second one is pulled to the opposite end of the pond (Figure 36b)

Ponds which are too wide to harvest by pulling each end of the seine down the long sides of the pond can also be seined but less efficiently so. The seining operation has to be repeated several times, keeping one end stationary and pulling the other end in a semicircular fashion (Figure 36c). Beating the surface of the water with sticks is often used as a method of discouraging the prawns from escaping the open end of the net as it is pulled toward the bank.

Generally speaking, the more men that are available to help with seining, the more efficiently this task can be performed (provided there is one clear leader). Three to five men can pull a seine net through a 30-m wide pond and seven to ten men can cope with a 50-m wide pond. To harvest a 0.8 ha pond by the two-seine technique mentioned above requires 12 to 15 men. A single-seined 0.2 ha pond takes about two to three hours to cull harvest using three or four men.

Seined prawns should be quickly transferred to a holding tank or net (Figure 40) from which they can be sorted. Market sized animals are retained for distribution to the market while smaller animals are returned to the original or another pond. Measurement of market size animals can be done by a quick check of total length (Figure 47). Tying the rulers to the operators hands is a useful way of doing this.

Circular "breeding" depressions are often seen in drained freshwater prawn ponds. These interfere with seining operations since large males can use them to evade the net.

8.5.2 Drain harvesting

The method and efficiency of drain harvesting depends on the design of the pond. As with any other method of harvesting, speed is important and harvesting should start very early in the morning while the temperature is cool. The level of the pond can be partially drawn down during the night before harvesting commences.

If the pond has a 'monk' or sluice gate structure for drainage, it is possible to include a harvesting sump in front of or beyond the gate or even contained within it (Figure 37). Alternatively, a temporary harvesting enclosure can be constructed (Figure 38). Further reading on the subject of harvesting boxes, sumps, etc., is suggested in Sivalingam (1974).

Most freshwater prawn ponds in Thailand (and in many other locations) do not have a 'monk' or a sluice gate and the pond must be drained by pumping. Prawns are prevented from entering the pump by means of a screen. This technique is illustrated in Figures 39 and 41. In this form of drain harvesting the prawns are caught by multiple seining of the pond while draining takes place. As the water level gets low, many prawns retreat into the mud or become stranded in isolated pools of water. At this stage there is no substitute for catching by hand (Figure 41). It is at this point that the farm operator wishes he had been more careful to construct his pond with a good slope towards the drain and a well compacted, smooth surface. The typical result of failure to do this, and the competitive 'harvesting' birds that stranded prawns attract, is well illustrated in Figure 42.

8.6 Post Harvest

What happens to freshwater prawns after harvest is a subject beyond the scope of this manual, but the topics of processing/marketing are dealt with in publications by Goodwin and Hanson (1975), Hanson and Goodwin (1977), and Shang *et al.* (1980).

Most freshwater prawns are sold close to the farm site, either on ice or alive. Prawns can be shipped in aerated water transport tanks to selected customers, such as hotels and restaurants, if the extra value for the live product warrants it.

Freshwater prawns are particularly subject to enzimatic damage after harvesting and death and some farms dip the prawns in iced water (kill-chill), blanch them by dipping them in water at 65°C for 15-20 seconds and then ice them and transport them to market. This technique needs a central processing area in the farm and the prawns must be brought to it live, in aerated containers.

(a) Sump before monk

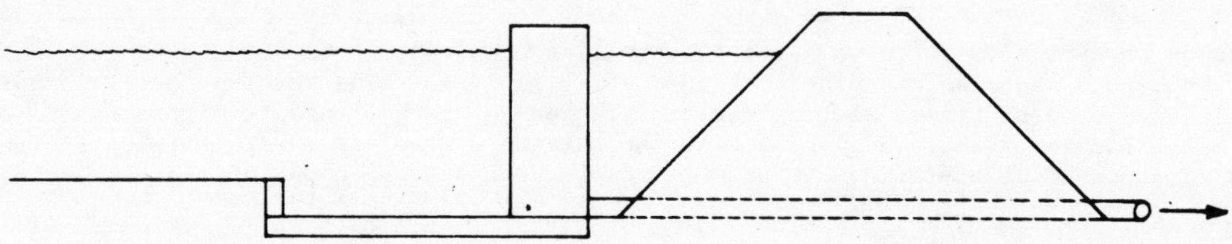

(b) Sump after monk

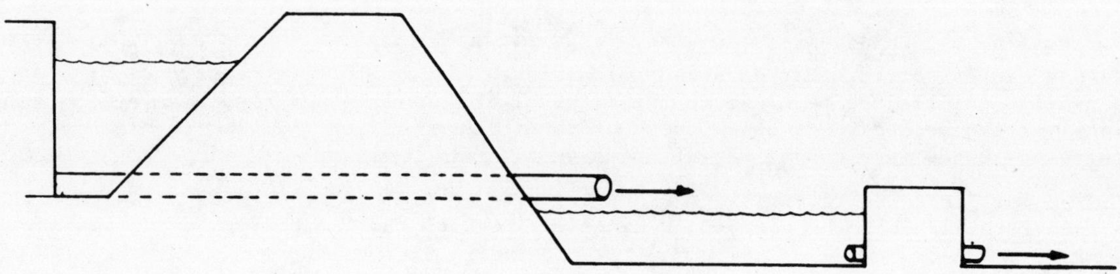

(c) Enclosure outside pond (see also Figure 38)

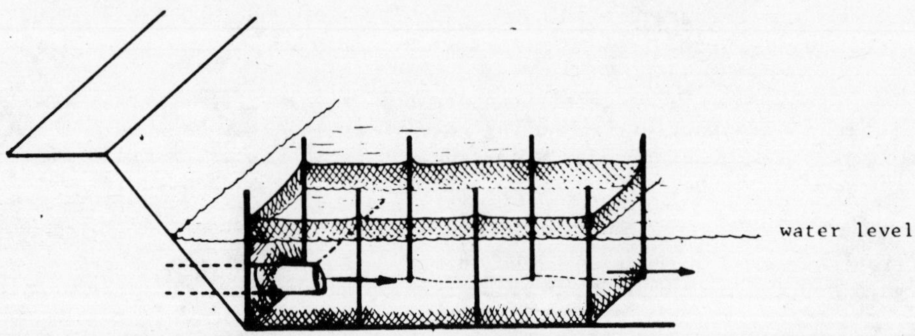

(d) Harvest area within sluice

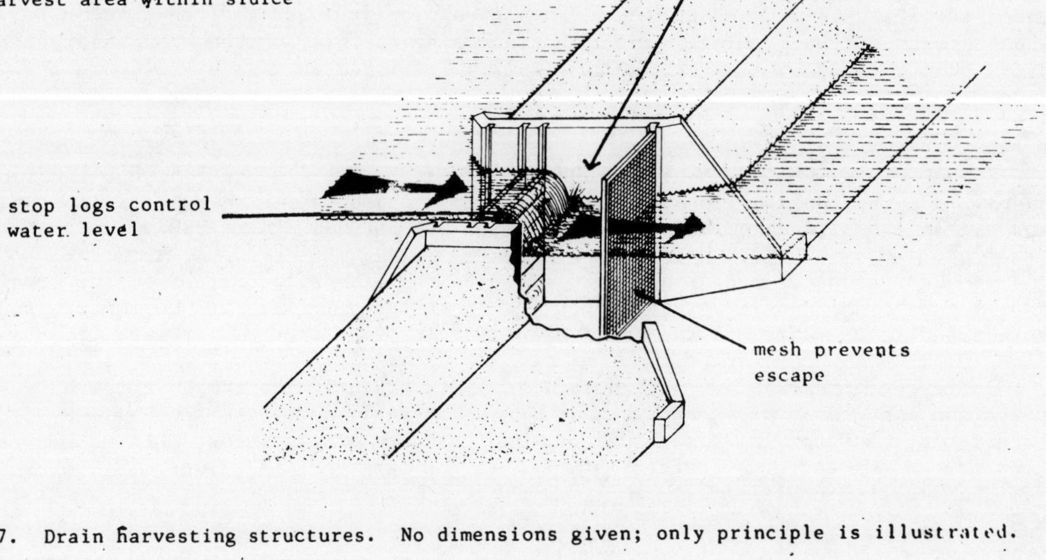

Figure 37. Drain harvesting structures. No dimensions given; only principle is illustrated.

Figure 38. Drain harvest net

Figure 39. Drain harvesting by pumping

Figure 40. Drain harvesting operations

Figure 41. Drain harvesting by pumping

Figure 42. The result of poor pond construction is at its most obvious at harvest time

Figure 43. Meat chopper used for pond diet preparation

Figure 44. Solar drier for pelleted food

Figure 45. Overhead view of a large freshwater prawn farm in Thailand

Generally the value of the harvested product will reflect its quality. Speed during and after harvesting, getting the prawns on ice and out of the sun and care in handling to prevent physical damage will all reap valuable dividends. The farmed prawn should be better than the wild caught product in every way and it is up to the farmer to see that his hard won harvest does not deteriorate through poor harvest and post-harvest procedures, ensuring that his prawns reach his customer absolutely fresh.

8.7 Problems

8.7.1 Predations

Predation is perhaps the greatest problem for any aquaculture enterprise, including freshwater prawn farming. Predation occurs mainly through other aquatic species, birds, snakes and humans. The use of small 60-cm high netting fences around ponds for the prevention of invading catfish and snakehead fish has already been mentioned (Figure 11, 16). The most troublesome predators (in Southeast Asia) are the snakehead (*Ophiocephalus micropites* and *O. striatus*), the catfish (*Clarias batrachus*) and the river catfish (*Mystus planiceps*). Crabs are a problem, especially as they bore holes in pond banks. They can be removed by sinking jars (traps) in the pond banks (Tunsutapanich, 1980b)

String is sometimes stretched across the top of ponds to deter birds. Bird-scaring devices, including explosive ones, are sometimes used and many farms regularly shoot invading birds.

Predation caused by other fish in the pond is controlled by adequate screening of the water intake. The ingress of fish eggs and larvae is not normally a problem in 'batch' culture of freshwater prawns since the fish grow at a similar rate than the prawns and sometimes form food for them. By the time they get to a dangerous size they are seined out during partial harvesting. To a certain extent this also happens in the 'continuous' culture technique but predation by 'top minnows', for example, on newly stocked post-larvae, can be a serious problem. The use of nursery ponds has been advocated (Wulff, 1982) as a means of preventing this loss. The presence of many frogs and toads in a pond usually indicates that predatory fish have been fairly efficiently excluded. In Taiwan (Liao and Liao, 1982) a dressing of 5 ppm of lime or tea seed cake (the residue, after oil extraction, from *Camellia sinensis*, which contains 10-13 percent saponin) is applied between 'batch' cycles to control unwanted fish in the ponds.

Undoubtedly the greatest source of loss in freshwater prawn farming is through human predation. Freshwater prawn farms are more prone to this type of predation than many fish farms because of the high value of the product and because it is relatively easy to catch. The temptation to catch a few kilogrammes of prawns by cast net at night (which are worth US$ 8.00/kg at the farm gate), when you are perhaps earning US$ 75.00/ month, is sometimes too great to resist. Human predation (poaching), like other forms of predation, can only be minimized by good management, not eliminated. Perimeter fences, dogs and reliable watchmen help. Big commercial farms which follow a policy of stocking some post-larvae into local public waters may also achieve some protection from predation, while the community will often protect cooperative-owned farms.

8.7.2 Water quality

This general topic has been dealt with earlier (Manual Sections 3 and 8.3).

Very high pH levels in freshwater prawn ponds are responsible for mortalities both because of the direct effect of the pH and the greater solubility of waste ammonia at high pH. High pH is often caused by a dense phytoplankton bloom. Successful control of these blooms with low levels (0.02 mg/litre) of an algicide (Clorosan) has been reported in freshwater prawn ponds (Aquacop, 1979b). pH levels fell rapidly and the algicide was not toxic to the prawns at this level.

The most likely source of external water pollution (not due to pond management practices) is from pesticides. Water levels of MIREX (a stable chlorinated hydrocarbon insecticide) of above 100 µg/litre are lethal, for example, while exposure to lower levels result in decreased productivity (Summer and Eversole, 1978). Thus the importance of site selection and water source may be demonstrated.

A scum of microscopic plants (phytoplankton) should not be allowed to cover the surface of the pond. It will cause low DO_2 problems at night and should be controlled by a reduction in feeding and by water exchange. Low DO_2 should be suspected if prawns begin to crawl out of the ponds or congregate at the edges of the pond in daylight. If this problem occurs, flushing is the solution.

8.7.3 Weeds

The presence of rooted plants in the pond, except those used to stabilize the pond banks, is a management problem. In fact, they may well be useful to the prawns, forming a source of food and a good habitat. However, their presence makes harvesting extremely difficult. They can be controlled in a number of ways:

(a) do not construct ponds with extensive shallow areas;

(b) never allow a pond to remain with only a small amount of water in it for long. 'Weeds' grow much better in shallow water (Figure 31);

(c) maintain an adequate phytoplankton bloom in the pond (Manual Section 8.3). This will reduce the light intensity at the bottom of the pond;

(d) cut the vegetation at emergence level. Pulling up the roots usually causes dangerous levels of turbidity in the pond. This job is very time- and labour-consuming and should not be necessary if the pond has been well constructed and managed.

8.7.4 Disease

'Shell disease' or 'black spot' which is the most obvious disease of post-larval, harvest-size prawns, is caused by an invasion of chitinolytic bacteria (bacteria able to break down the chitin of the exoskeleton) and sometimes, later, by a fungus. This disease does cause mortalities and also reduces the value of harvested prawns through disfigurement. It follows physical damage which, in itself, is often caused by aggression between the prawns. Careful handling and the reduction of stocking levels usually minimizes the incidence of this problem.

Filamentous bacteria, particularly when in the gill chambers, also cause problems, particularly when animals are crowded (as they are in a holding tank for post-larvae or a nursery pond).

Opacity or 'whiteness' of the muscular tissues, which often proceed forward from the tail, seems to be a reaction to stress conditions such as overcrowding, low oxigen levels, temperatures outside the optimum range or temperature shock, pH, etc.; often the condition seems to be reversible, not necessarily fatal.

Johnson (1982) considers that the dark areas often noted in the gill chamber are due to precipitating chemicals and high levels of nitrogenous wastes.

Inability to moult successfully can be caused by poor water quality and the presence of toxins but also by nutritional deficiency.

The control of disease problems in the pond culture of freshwater prawns is a subject which has had little attention to date, perhaps because diseases at this growth stage have not been a major problem. As densities are pushed up and farmers try to increase productivity, disease problems are bound to become more prevalent. Further reading on prawn diseases may be found in Sindermann (1977), Johnson (1977, 1982), Hanson and Goodwin (1977) and Goodwin and Hanson (1975).

8.8 Alternative Rearing Techniques

The following sections of the manual deal only briefly with other methods of rearing prawns to market size but, where possible, give references for further reading.

8.8.1 Nursery ponds

Some farmers use nursery ponds into which newly metamorphosed post-larvae are stocked and reared at much higher densities than discussed in Manual Section 8.2. The animals are grown in the nursery ponds for one or two months before harvesting and stocking into production ponds.

This technique is used for one of at least four reasons. One is to produce larger juveniles to stock in a rearing pond (sometimes especially applicable in an area with a limited growing season; in this case the post-larvae may be reared on a different site or even, where it is temperature which limits the growing season, indoors under temperature control). Another reason is to achieve better survival rates while the post-larvae are particularly susceptible to predation. Predation is easier to prevent in a 'batch' than a 'continuous' system of pond management. Post-larvae are therefore sometimes grown in 'batch' operated nursery ponds before stocking at a larger size into 'continuous' production ponds. Thirdly, higher stocking densities result in a more efficient use of food in nursery ponds than when young post-larvae are stocked directly into production ponds. The fourth reason is that of accountability. As noted in Appendix 6, post-larvae are particularly difficult to count accurately. It is much easier to count larger juveniles, after initial growth in a nursery pond. Additionally, poor survival rate of a particular batch is easier to notice in a crowded nursery pond than it is in a production pond. It is much cheaper to replace a poor batch of post-larvae at this stage than not to realize the problem until harvest time. The result of employing a nursery pond is therefore the use of better quality animals for the subsequent production ponds and greater accuracy in stocking density. The use of nursery ponds was suggested by Ling (1969) many years ago.

Using this technique, Wulff (1982), for example, obtains a nursery survival rate of 70 percent. Post-larvae are stocked into the nursery ponds at a density of 1 000/m^2 and grown there for 2 1/2 months before being counted and transferred to production ponds. Smith and Sandifer (1979) describe the use of indoor nursery tanks on recirculating water systems stocked at densities between 300/m^2 and 1 500/m^2.

8.8.2 Cage and pen culture

The use of these two types of structures (a cage is defined as a floating structure enclosed usually in netting, while a pen is an area of a larger water body, such as a reservoir or lake, which is separated off by the use of a netting, bamboo or other structure) has not yet been practised in freshwater prawn culture to any degree. However, small-scale cage and pen experiments have taken place in Thailand and initial results in Songkhla Lake with pen culture have encouraged the Asian Development Bank to support the extension of this technique there (International Agro-Fisheries System, 1978). The use of nylon cages suspended in irrigation ditches has been reported by Menasveta and Piyatiratitivokul (1982).

8.8.3 Polyculture

The culture of other aquatic species with freshwater prawns has also not gained much prominence yet although it has been practised in experimental work by Tunsutapanich *et al.* (1982), in Thailand, who used grass, silver and bighead carp. Mullet was cultured with *M. acanthurus* by Martinez *et al.* (1977), while Stickney (1980) and Buck *et al.* (1979) report polyculture experiment with tilapia and Chinese carps, respectively. Bighead and grass carp are used by farmers in Taiwan, where polyculture experiments of freshwater prawns with mullet and milkfish are also reported (Liao and Liao, 1982).

8.8.4 Tank culture

Many attempts have been made to grow freshwater prawns under highly intensive conditions in tanks housed under environmentally controlled conditions in temperate zones. Such ideas have generally been abandoned except for systems for culturing prawns in recirculating systems in greenhouses being marketed in the USA (Anon., 1980) and experimental work on the use of thermal effluents and geothermal water.

Sandifer *et al.* (1982) have reported that the prospects for the success of semi-intensive culture of freshwater prawns in tanks using artificial habitats to increase surface area, aeration, etc., are good. Initial work in 173 m^2 concrete tanks stocked at 83 and 32 post-larvae/m^2 gave production rates equivalent to 3 800-4 700 kg/ha/year.

The distinctive feature about tank culture is the absence of soil nutrients and a marked reduction in the amount of natural food available. Despite this, Boonyaratpalin and New (1982) reported production rates of 465-820 kg/ha in 119 days through culture of freshwater prawns in 50 m^2 concrete tanks stocked at $5/m^2$. Where available, it seems that concrete (reservoir) tanks may be successfully used to grow freshwater prawns, at least at low densities.

8.8.5 Open water stocking

Freshwater prawns may also be stocked into reservoirs, rivers and lakes. This is particularly valuable, especially in areas where *Macrobrachium* was formerly abundant but the fishery has declined or disappeared due to overfishing and physical interference (dams) with the completion of its life cycle. Stocking into natural bodies of water has been quite effective in Thailand as a source of food and revenue for local fishermen.

Before commencing the restocking of open waters where *Macrobrachium* were formerly abundant, it is essential to determine the cause of the decline in the natural fishery. If this has been caused by agricultural development in the watershed, with resultant pollution with pesticides, a restocking policy will be of little avail.

9. ACKNOWLEDGEMENTS

The authors would like to thank the Fisheries Department of the Food and Agriculture Organization of the United Nations for suggesting and sponsoring the preparation of this manual. We also thank the Department of Fisheries of the Ministry of Agriculture and Cooperatives, Royal Thai Government, for their support and encouragement.

Our personal thanks go to all the staff of the Chacheongsao Fisheries Station, whose hard work and experience has made it easier to prepare this document. We are most grateful to Mario Pedini, who reviewed the draft of the manual, provided some photographs, many useful comments and suggestions, and edited it, and to Paolo Lastrico, who kindly prepared the diagrams. Dr. A. Coche of the FAO Fisheries Department is also thanked for providing some photographs.

Finally, we thank a pioneer of freshwater prawn farming, Takuji Fujimura, who has been a personal inspiration to us both. Fuji very kindly proofread this Manual and provided constructive criticism and information. We are particularly grateful for permission to use his beautiful transparencies of the larval stages of <u>Macrobachium</u> <u>rosenbergii</u>.

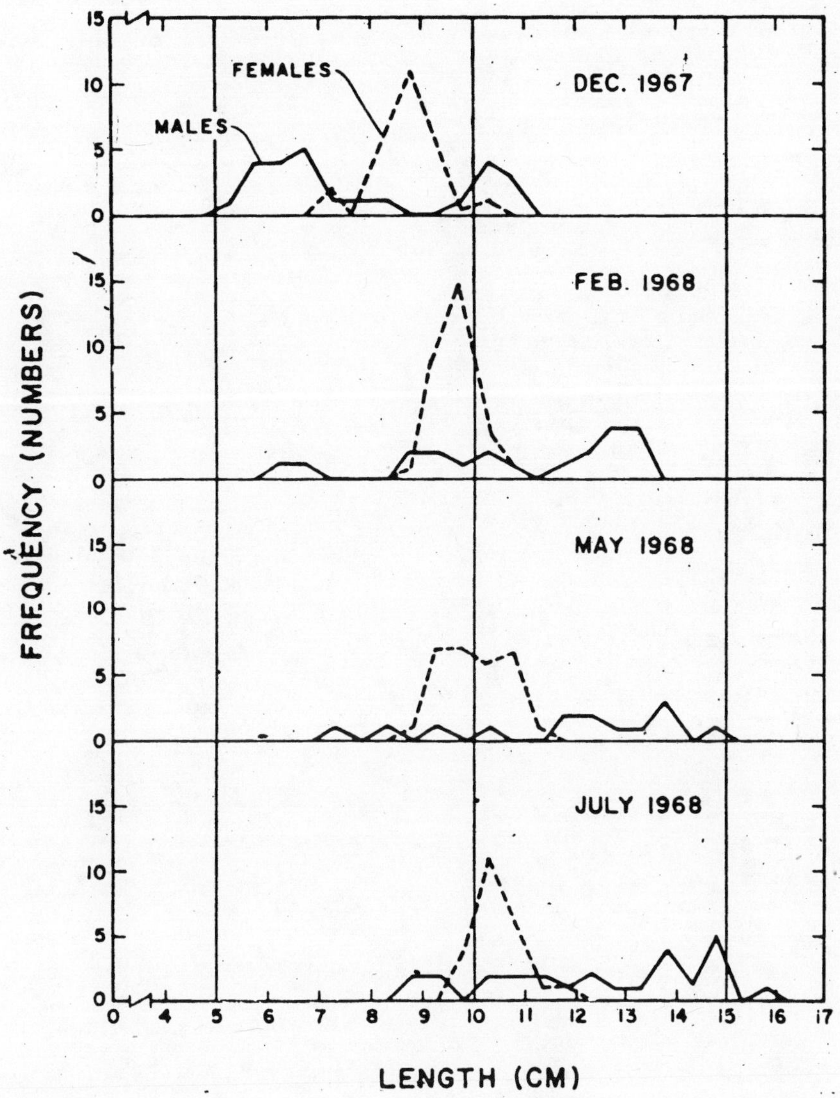

Figure 46. Length frequency distribution of <u>Macrobrachium rosenbergii</u> population (Fujimura, 1974).
Length frequency distributions of males and females from an unharvested population reared in an earthen pond (Quarter House Ranch, Kanai, Hawaii)

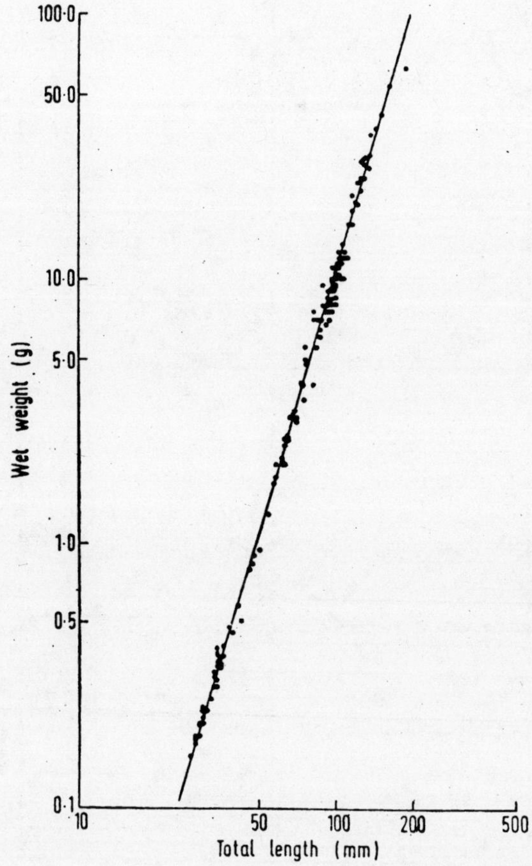

Figure 47. Length/weight relationship of Macrobrachium rosenbergii (Wickins, 1972)

10. REFERENCES

AQUACOP, 1977 Macrobrachium rosenbergii culture in Polynesia: progress in developing a mass intensive larval rearing technique in clear water. Proc.World Maricult.Soc., 8:311-26

__________, 1979 Intensive larval culture of Macrobrachium rosenbergii: a cost study. Proc.World Maricult.Soc., 10:429-34

__________, 1979a Macrobrachium rosenbergii culture in Polynesia: pH control in experimental pond waters by phytoplankton limitation with an algicide. Proc.World Maricult.Soc., 10:392-402

Armstrong, D.A., M.J. Stephenson and A.W. Knight, 1976 Acute toxicity of nitrite to larvae of the giant Malaysian prawn, Macrobrachium rosenbergii. Aquaculture, 9:39-46

Balazs, G.H. and E. Ross, 1976 Effect of protein source and level on growth and performance of captive freshwater prawn, Macrobrachium rosenbergii. Aquaculture, 7:299-313

Balazs, G.H., E. Ross and C.C. Brooks, 1973 Preliminary studies on the preparation and feeding of crustacean diets. Aquaculture, 2:369-77

Bardach, J.E., J.H. Ryther and W.O. McLarney, 1972 Aquaculture: the farming and husbandry of freshwater and marine organisms. New York, Wiley-Interscience, U.S.A.

Barnes, A., 1982 (Research discussion Session held during Giant Prawn 1980, Bangkok, Thailand, June 15-21, 1980). In Giant prawn farming, edited by M.B. New. Amsterdam, Elsevier

Bartlett, P. and E. Enkerlin, 1983 Growth of the prawn Macrobrachium rosenbergii in asbestos asphalt ponds in hard water on a low protein diet. Aquaculture, 30:353-6

Boonyaratpalin, M. and M.B. New, 1982 Evaluation of diets for Macrobrachium rosenbergii reared in concrete ponds. In Giant prawn farming, edited by M.B. New. Amsterdam, Elsevier, pp. 249-56

Boyd, C.E., 1979 Water quality in warmwater fish ponds. Auburn, Alabama, Auburn University Agricultural Experiment Station

Buck, D.H., et al., 1979 Polyculture of Macrobrachium rosenbergii and Chinese carps in ponds enriched with swine manure. Paper presented at the 10th Annual Meeting of the World Mariculture Society. Honolulu, Hawaii, January 22-26, 1979

Cansdale, G.S., 1979 Application of low-cost filtration systems to freshwater prawn farming in Thailand. UNDP/FAO Programme for the Expansion of Freshwater Prawn Farming Working Paper. Rome, FAO, THA/75/008/79/WP/8

Ferdinando, R.L. and S.A.W. Manawadu, 1982 Some experiences of a commercial Macrobrachium farm in Sri Lanka. In Giant prawn farming, edited by M.B. New. Amsterdam, Elsevier, pp. 441-4

Fincham, A.A. and J.F. Wickins, 1976 Identification of commercial prawns and shrimps. Brit.Mus.Publ., (779):7 p.

Forster, J.R.M. and J.F. Wickins, 1972 Prawn culture in the United Kingdom: its status and potential. Lab.Leafl.Minist.Agric.Fish.Food G.B. (New Ser.), (27):32 p.

Fujimura, T., 1974 Development of a prawn culture industry in Hawaii, Job Completion Report for Project H-14-D, (period from 1 July 1969 - 30 June 1972), Department of Land and Natural Resources, State of Hawaii (internal report)

__________, 1978 Plan for the development of prawn farming in Thailand and recommendations to increase production of juveniles for distribution to farmers and for stocking natural areas. UNDP/FAO Programme for the Expansion of Freshwater Prawn Farming Working Paper. Rome, FAO, THA/75/008/79/WP/2

Fujimura, T. and H. Okamoto, Notes on progress made in developing a mass culturing technique for
1972 Macrobrachium rosenbergii in Hawaii. In Coastal aquaculture in the Indo-Pacific Region, edited by T.V.R. Pillay. West Byfleet, England, Fishing News Books Ltd., for IPFC/FAO, pp. 313-27

Goodwin, H.L. and J.A. Hanson, The aquaculture of freshwater prawns (Macrobrachium species).
1975 Waimanalo, Hawaii, The Oceanic Institute

Hanson, J.A. and H.L. Goodwin (eds), Shrimp and prawn farming in the western hemisphere.
1977 Stroudsburg, Pennsylvania, Dowden, Hutchison and Ross, 439 p.

Holthuis, L.A., FAO species catalogue. Vol. I. Shrimps and prawns of the world (an annotated
1980 catalogue of species of interest to fisheries). FAO Fish.Synop., (125)vol.I:261 p.

International Agro-Fisheries System, Inland fisheries development project of the Kingdom of
1978 Thailand, Final report, Volume 2. Manila, Philippines, Asian Development Bank

Jamandre, T.J., Pumps for brackishwater aquaculture. In Joint SCSP/SEAFDEC Workshop on
1977 Aquaculture Engineering. Volume 2. Technical report. Manila, Philippines, South China Sea Fisheries Development and Coordinating Programme, (SCS/GEN/77/15): 393-422

Johnson, S.K., Crawfish and freshwater shrimp diseases. College Station, Texas, Texas A & M
1977 University, (TAMU-SC-77-605)

__________, Diseases of Macrobrachium. In Giant prawn farming, edited by M.B. New.
1982 Amsterdam, Elsevier, pp. 269-77

Liao, I-C., and N-H. Liao, Progress of Macrobrachium farming and its extension in Taiwan. In Giant
1982 prawn farming, edited by M.B. New. Amsterdam, Elsevier, pp. 357-79

Ling, S-W., The general biology and development of Macrobrachium rosenbergii. FAO Fish.Rep.,
1969 (57)vol.3:589-606

__________, Methods of rearing and culturing Macrobrachium rosenbergii. FAO Fish.Rep.,
1969a (57)vol.3: 607-19

Martinez-Silva, L.E., M. Pedini and M.B. New, Mullet (Mugil incilis) and freshwater prawn
1977 (Macrobrachium rosenbergii) polyculture in Colombia. Proc.World Maricult.Soc., 8:195-206

Menasveta, P., Effect of ozone treatment on the survival of prawn larvae (Macrobrachium
1982 rosenbergii) reared in a closed-recirculating water system. In Giant prawn farming, edited by M.B. New. Amsterdam, Elsevier, pp. 295-302

Menasveta, P. and S. Piyatiratitvokul, Effects of different culture systems on growth survival and
1982 production of the giant freshwater prawn (Macrobrachium rosenbergii). In Giant prawn farming, edited by M.B. New. Amsterdam, Elsevier, pp. 175-89

New, M.B., The selection of sites for aquaculture. Proc.World Maricult.Soc., 6:379-88
1975

__________, A review of dietary studies with shrimp and prawns. Aquaculture, 9:101-44
1976

__________, The diet of prawns. UNDP/FAO Programme for the Expansion of Freshwater Prawn
1980 Farming Working Paper. Rome, FAO, THA/75/008/80/WP/12

__________, A bibliography of shrimp and prawn nutrition. Aquaculture, 21:101-28
1980a

__________, (ed.), Giant prawn farming. Amsterdam, Elsevier, 532 p.
1982

New, M.B., S. Singholka and P. Vorasayan, Current status of freshwater prawn farming in Thailand.
1982 In Giant prawn farming, edited by M.B. New. Amsterdam, Elsevier, pp. 333-49

New, M.B., et al., A recirculation system for experimental aquaria. Aquaculture, 3:95-103
1974

__________, Freshwater prawn farming in Dominica: a feasibility report. Prepared for the
1977 Governments of Dominica and U.K. Marlow, Bucks., England, Kelvin Hughes Aquaculture Services

Popper, D.M. and R. Davidson, An experiment in rearing freshwater prawns in brackish water. In
1982 Giant prawn farming, edited by M.B. New. Amsterdam, Elsevier, p. 173

Sandifer, P.A., Semi-intensive grow-out of prawns (Macrobrachium rosenbergii): preliminary results
1982 and prospects. In Giant prawn farming, edited by M.B. New. Amsterdam, Elsevier, pp. 161-72

Sarver, D., S. Malecha and D. Onizuka, Possible sources of variability in stocking mortality in
1982 postlarval Macrobrachium rosenbergii. In Giant prawn farming, edited by M.B. New. Amsterdam, Elsevier, pp. 99-113

Shang, Y.C., et al., Thai freshwater prawn and brine shrimp farming: report on a study of economics,
1980 marketing and processing requirements. UNDP/FAO Programme for the Expansion of Freshwater Prawn Farming Working Paper. Rome, FAO, THA/75/008/80/WP/17

Sindermann, C.J., Freshwater shrimp diseases. General. In Disease diagnosis and control in North
1977 American marine aquaculture, edited by C.J. Sindermann. Dev.Aquacult.Fish.Sci., 6: 78-95

Singholka, S., Observations on the design, construction and management of small-scale or backyard
1978 hatcheries for Macrobrachium rosenbergii in Thailand. UNDP/FAO Programme for the Expansion of Freshwater Prawn Farming Working Paper. Rome, FAO, THA/75/008/78/WP/1

Singholka, S. and C. Sukapunt, Use of a simple recirculation system for larval culture of Macro-
1982 brachium rosenbergii. In Giant prawn farming, edited by M.B. New. Amsterdam, Elsevier, pp. 291-3

Sivalingam, S., A guide to construction of fish ponds. Fed.Fish.Occas.Pap.Fed.Dep.Fish.Niger.,
1974 (20):64 p.

Smith, T.I.J., P.A. Sandifer and W.E. Jenkins, Growth and survival of prawns, Macrobrachium
1982 rosenbergii, pond reared at different salinities. In Giant prawn farming, edited by M.B. New. Amsterdam, Elsevier, pp. 191-202

Sorgeloos, P., The culture and use of brine shrimp, Artemia salina, as food for hatchery raised larval
1978 prawns, shrimps and fish in South-East Asia. UNDP/FAO Programme for the Expansion of Freshwater Prawn Farming Working Paper Rome, FAO, THA/75/008/78/WP/3

Sorgeloos, P., et al., Decapsulation of Artemia cysts: a simple technique for the improvement of the
1977 use of brine shrimp in aquaculture. Aquaculture, 12:311-5

Spotte, S., Fish and invertebrate culture: water management in closed systems. New York, Wiley-
1970 Interscience, 145 p.

Stahl, M.S., The role of natural productivity and applied feeds in the growth of Macrobrachium
1979 rosenbergii. Proc.World Maricult.Soc., 10:92-109

Stickney, R.R., Options for tilapia producers - bait to broilers. Paper presented to the Texas Fish
1980 Farming Conference, College Station, Texas, January 23, 1980

Strickland, J.D.H. and T.R. Parsons, A practical handbook of seawater analysis. Bull.Fish.Res.Board
1968 Can., (167):311 p. Rev.ed. of Bull.Fish.Res.Board Can., (125)

Suharto, H.H., A. Ismail and A. Poernomo, Breeding technique of Macrobrachium rosenbergii in conical fibreglass tanks. In Giant prawn farming, edited by M.B. New. Amsterdam, Elsevier, pp. 115-22
1982

Summer, S.E. and A.G. Eversole, Effects of water-borne mirex on the survival and production of Macrobrachium rosenbergii. Proc.World Maricult.Soc., 9:47-54
1978

Suwannatous, S. and M.B. New, Development of a simple shallow beach well for hatcheries requiring sea water. In Giant prawn farming, edited by M.B. New. Amsterdam, Elsevier, pp. 303-7
1982

Terazaki, M., P. Tharmbuppa and Y. Nakayama, Eradication of predatory fishes in shrimp farms by utilization of Thai tea seed. Aquaculture, 19:235-42
1980

Tunsutapanich, A., An improved technique for decapsulation and preservation of Artemia cysts (brine shrimp eggs) developed at the Chacheongsao Fisheries Station. UNDP/FAO Programme for the Expansion of Freshwater Prawn Farming Working Paper. Rome, FAO, THA/75/008/79/WP/6
1979

__________, The use of rock salt, sea salt, rock salt brine, and salt stock solution for larval culture of Macrobrachium rosenbergii. UNDP/FAO Programme for the Expansion of Freshwater Prawn Farming Working Paper. Rome, FAO, THA/75/008/80/WP/18
1980

__________, The eradication of crabs in Macrobrachium ponds. In: Giant prawn farming, edited by M.B. New. Amsterdam, Elsevier, pp. 279-87
1982

Tunsutapanich, A., S. Chalaypote and P. Phuhoung, Macrobrachium farming in areas with irregular water supply. In Giant prawn farming, edited by M.B. New. Amsterdam, Elsevier, pp. 207-12
1982

Uno, Y. and K.C. Soo, Larval development of Macrobrachium rosenbergii reared in the laboratory. J.Tokyo Univ.Fish., 55(2):179-90
1969

Vanhaecke, P. and P. Sorgeloos, International study on Artemia. 19: Hatching data on 10 commercial sources of brine shrimp cysts and re-evaluation of the "hatching efficiency" concept. Proc.World Maricult.Soc., 12 (in press)

Webber, H.H., Risks to the aquaculture enterprise. Proc.World Maricult.Soc., 4:19-33
1973

Weidenbach, R.P., Dietary components of freshwater prawns reared in Hawaian ponds. In Giant prawn farming, edited by M.B. New. Amsterdam, Elsevier, pp. 257-67
1982

Wheaton, F.W., Aquacultural engineering. New York, Wiley-Interscience, 708 p.
1977

Wickins, J.F., Experiments on the culture of the spot prawn Pandalus platyceros and the giant freshwater prawn Macrobrachium rosenbergii. Fish.Invest.Minist.Agric.Fish.Food G.B. (2 Sea Fish.), 27(5): 23 p.
1972

__________, The tolerance of warm-water prawns to recirculated water. Aquaculture, 9:19-37
1976

Willis, S.A. and M.E. Berrigan, Effect of stocking size and density on growth and survival of Macrobrachium rosenbergii in ponds. Proc.World Maricult.Soc., 8:251-64
1977

Wulff, R.E., The experience of a freshwater prawn farm in Honduras, Central America. In Giant prawn farming, edited by M.B. New. Amsterdam, Elsevier, pp. 445-8
1982

Anon., News item about "The Shrimpery". In Aquaculture Digest, edited by R. Rosenberry. San Diego, California, U.S.A.
1980

Table 1

Typical analysis of fresh water at two successful freshwater prawn hatcheries[1/]

Location:	Honolulu, Hawaii	Vairao, Tahiti
Type:	Tap water	Small mountain stream
Date analyzed:	19 October 1971	29 September 1976
pH	8.0	7.7
Free CO_2	nil	NA
Total hardness (as $CaCO_3$)	71	54
Ca	12	8
Mg	9.9	8.2
Cl (as NaCl)	81	10
Fe	<0.02	nil
SO_4	7.9	3.0
SiO_2	41	21
PO_4	0.15	nil
Free NH_3-N	NA	nil
NO_2-N	NA	nil
NO_3-N	NA	0.55
Na	28	6
K	2.0	0.6
TDS	217	NA
Turbidity (J.T.U.)	nil	nil

1/ Except for pH, data are given in ppm (parts per million)

NA not analyzed

Table 2

Example of quantity of sea-water needed to make brackish-water for larval culture

Sea-water salinity (°/oo)	Amounts of water required to make 10 m^3 of 12‰ brackish-water	
	Fresh-water[1/] (m^3)	Sea-water (m^3)
36	6.67	3.33
35	6.57	3.43
34	6.47	3.53
33	6.36	3.64
32	6.25	3.75
31	6.13	3.87
30	6.00	4.00
29	5.86	4.14
28	5.71	4.29
27	5.56	4.44
26	5.38	4.62
25	5.20	4.80
24	5.00	5.00

1/ Assumed to be 0°/oo salinity

Table 3

Relationship between temperature, salinity and dissolved oxygen (in ppm)

The following table has been adopted from Spotte (1970) using Knudsen's formula for converting chlorinity to salinity

Temperature (°C)	Salinity (°/oo)				
	0	7.5	11.1	14.7	36.4
20	9.1	8.7	8.5	8.3	7.4
22	8.8	8.4	8.2	8.0	7.1
24	8.4	8.1	7.9	7.7	6.9
26	8.1	7.8	7.6	7.5	6.6
27	8.0	7.6	7.5	7.3	6.5
28	7.8	7.5	7.4	7.2	6.4
29	7.7	7.4	7.2	7.1	6.3
30	7.6	7.3	7.1	7.0	6.2

Table 4

Hatchery feeding schedule[1/]

Larval age (days)	Brine shrimp nauplii (BSN)		Prepared feed (PF)	Pond feed or floating diet
	Morning feed	Evening feed		
1	–	✓	–	–
2	✓	✓	–	–
3	✓	✓	✓	–
4	✓	✓	✓	–
5	–	✓	✓	–
6-(metamorphosis)	–	✓	✓	–
pl onward	–	–	–	

1/ The quantities of brine shrimp nauplii and prepared feed will vary according to demand but will gradually increase as the larvae grow. Exact daily quantities depend on animal demand; this topic is fully discussed in Manual Section 5.3

Table 5

Water requirements for rearing freshwater prawns in ponds

Total farm water surface area[1/] (ha)	Quantity of water required			
	Pond filling[2/] (m^3/min) Maximum	Maintenance flow[3/] (m^3/min) Minimum	Maintenance flow[3/] (m^3/min) Maximum	Consumption[4/] (m^3/min) Average
0.2	2.50	0.03	0.11	0.12
0.5	2.50	0.07	0.28	0.31
1.0	2.50	0.14	0.56	0.68
2.0	2.50	0.28	1.12	1.26
3.0	3.75	0.42	1.68	1.89
5.0	6.25	0.7	2.8	3.14
10.0	12.5	1.4	5.6	6.28
20.0	25.0	2.8	11.2	12.57
40.0	50.0	5.6	22.4	25.14

1/ Average water depth 0.9 m

2/ For filling ponds initially and subsequently and for rapid flushing in emergency. Assumes (a) that the unit pond size is 0.2 ha and that a pond can be filled within 12 hours, and (b) that it will never be necessary to fill more than one pond (or 10 percent of the farm surface area, whichever is the greater) at one time. Local experience will tell if this allowance is either insufficient or too generous

3/ Continuous water demand, based on 140 l/ha/min minimum and 560 l/ha/min maximum. Actual requirement depends on whether "continuous" culture is practised or not. The minimum figure should cope with average seepage and evaporation losses

4/ This consists of the maximum maintenance rate, assuming "continuous" culture for the whole farm, plus the quantity necessary to fill all ponds four times per year, averaged out to a volume per minute average consumption basis

Table 6

Average discharge from outlet pipes[1/]

Pipe diameter (m)	Pipe discharge (m^3/min)
0.3	4.8
0.4	7.8
0.5	12
0.6	18
0.8	30
1.0	48

1/ These figures are approximate. Actual discharge depends on the head of water above the pipe. In this case a head of 1.5 m is assumed

APPENDIX 1

Water Filtration

The following text is a reproduction of some notes by Cansdale (1979) on methods for low-cost water filtration for freshwater prawn farms.

A simple screen well, made of plastic and costing US$ 10 in 1979, which was based on the 'Cansdale' technique, has been described by Suwannatous and New (1982). Vertical and, where the beach is too shallow over rock for this, horizontal versions of this point well have been in successful use in Thailand since 1979. Its operational principle is the same as that of the units described here. The units described by Cansdale are more adaptable to sites where the point screen well is inapplicable; they can also be used in artificial media, as explained in this Appendix.

1. INSTALLATION AND USE OF FILTER UNITS

The SWS is easy to install and run. Once the principle is understood, it can be used successfully in a range of situations and for many purposes, but it is not a magic recipe for instant-free clean water. The advice in this Appendix must be studied carefully, giving special attention to:

- choice of site, pump and pipeline, and
- correct installation and thorough development.

These instructions should allow most sites to be worked but there is great advantage in having a small trained mobile team to establish units and teach other groups. This basic training can be given by SWS in Britain or overseas. It is possible for an experienced person to work alone but better to have a team of two or three, of which one has some experience of machines. Technical advice should be asked about difficult sites and also about projects where large volumes are needed.

1.1 General Principles

The SWS Unit is not itself a filter but a device for making the sea or river bed serve as a natural sand filter. Dirt is held on the bed surface as water is drawn to the open bottom of the unit through an area of up to 10-m diameter. Water moving over the bed cleans it, helped by fish, etc., that poke about and disturb the sediment. In the sea the action of waves and tides is effective. The bed itself becomes a biological filter that destroys anaerobic bacteria and reduces the levels of ammonia, iron, BOD, etc. (Figure 1)

1.2 Site Requirements

A permeable bed is needed at least 60-cm deep, but preferably not less than 1 m, and under a minimum of 30-cm water. Different types of site, including those to which sand must be added, are described in the following pages. Beds of a wide range may be used - sand, gravel, broken coral, shell, etc. The bulk of the grains should be between 0.5 mm and 5.0 mm (0.02-0.2 in) but a great advantage of this system is that during development, described below, excess fine sand is pumped out, leaving the larger grains in and around the unit so a precise sand specification is not needed. Uniformly fine sand, especially of wind-blown origin, is unsuitable on its own, but it can be graded up by adding coarse sand or gravel under and around the unit. If most grains are above 2 mm (0.1 in) diameter, it helps to add fine sand on the surface around the unit during development. "Fine sand" is material up to 1 mm and "coarse sand" from 2 mm to 5 mm, but these are not used as technical terms. A few stones up to 50 mm (2 in) do little harm but larger stones reduce the bed's efficiency and should be cleared around the unit.

1.3 Installations

Dig a hole with centre deep enough to take the unit, open end down, so that its top is up to 15 cm (6 in) buried when stabilized. Sand is filled in around the unit and more piled over to allow for settlement. If a hollow forms, more sand should be drawn over it.

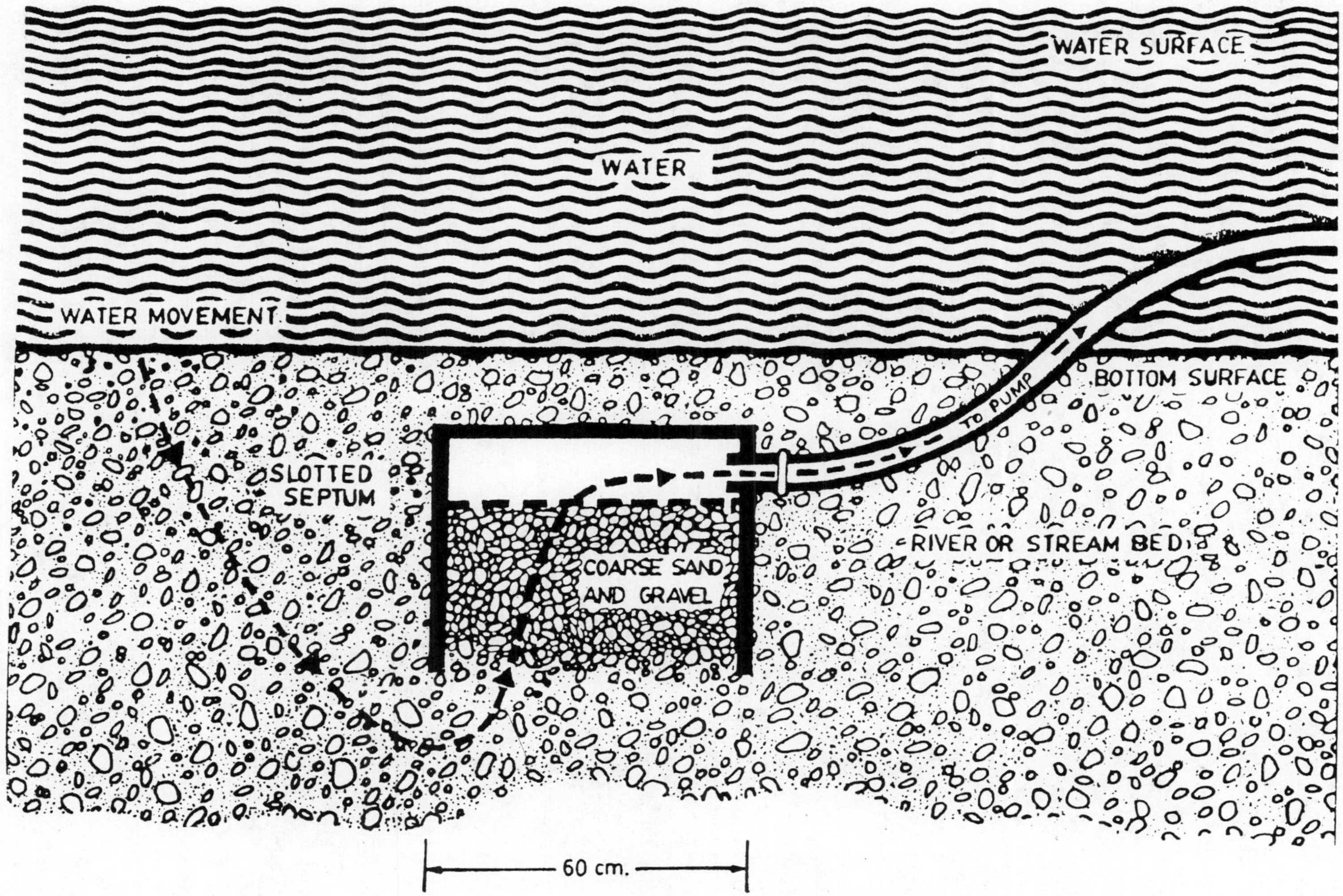

Figure 1. Diagram of SWS box unit and water flow

A special tool is supplied with each unit, or set of units, for digging this hole. A 2-m pole is fastened to it and it is used like a rake.

In cold climates, most work can be done in waters but sometimes a skin-diver is needed, especially in the sea. Men can work comfortably in warm water, but special precautions must be taken in bilharzia areas.

1.4 Filter Development

Thorough development is the key to success and this section is most important.

When the unit is buried and the suction line has filled with water, this is then connected to the pump intake. Tight joints with washers are essential, for the smallest air leak delays priming and lowers efficiency. Under-water leaks may admit raw water though, if these are very small, they soon block. Develop with a temporary pump close to the water or work from a boat, especially if the permanent pump is to be submersible. It may help to stand on the unit, rocking it gently to settle it until the pump primes fully: the speed should then be reduced until it runs steadily. At first the water is full of silt and organic matter as it cleans the bed. Varying with the site, this water clears in anything from a few to many minutes. The pump is then stopped and restarted; after a very short interval, the water becomes dirty but soon clears, when the pump is again stopped and restarted. Releasing the partial vacuum disturbs the sand in and around the unit, allowing more fine material to be sucked out and gradually pushing back the perimeter of clean coarse sand, so improving flow to the unit, which is one of the basic reasons for development. This process continues until water no longer becomes dirty after restarting, and the pump is working to full capacity. The type of bed and the pump size determine whether this takes an hour or perhaps a whole day.

Where the bed has much black organic matter, development is best spread over several days to allow this to decay aerobically, after which it is easily sucked out.

The water should now be crystal clear, free of all suspended matter and organisms down to about 1 micron and sometimes below that. It is suitable for most purposes, including for village supplies where the water had previously been used raw. Where sea water is wanted for research or fresh water for town supply, it should be pumped to waste several hours daily for at least a week while the biological filter develops in the bed, the time needed varying with temperature and other factors. Where quality is critical, this should be monitored.

Where adverse site conditions impede progress, the following procedures may be tried:

(a) using a garden fork or similar tool, dig the bed well around the unit, letting the stream carry away much of the silt;

(b) instead of just stopping the pump, release the suction completely, letting the water flow back to the unit, before re-starting;

(c) interchange intake and take-off hoses and pump back for several minutes. Somebody should stand on the unit until normal pumping restarts.

Small amounts of sand may be drawn through for some days, especially when pumping is periodic, but this is sterile and settles at once in a reservoir or small baffle chamber

If pump and pipeline are not limiting factors, the rate of flow is largely determined by the site and large volume is often possible. For drinking water it is better to underpump and a steady flow of not above 20 m^3/h (4 500 gal) per unit is suggested. Although the system is designed for continuous working in both sea and river, it is equally efficient when pumped periodically, preferably daily, but the system has been shown to be clean and effective in a British river after being left for three months. When the pump is started again, it is probably advisable to pump some water to waste; this need be for only a few minutes after a day's gap but perhaps for up to an hour after a week. Only local experience will show what is needed and in fact whether this is even necessary.

1.5 Sites with Ample Sand

1.5.1 Rivers and streams

Conditions vary with climate, shape of ground, geology, etc., so that the site must be chosen carefully. It must be stable and free from scour, which prevents the formation of a good filter bed and may even wash out the unit. The inside of a bend is often suitable. There should always be at least 30 cm water over it. In rivers which fluctuate widely the only permanent source is one covered at low water but, as the water rises, it may be necessary to take up the unit and replace it near the edge.

In rivers which stop flowing in the dry season and then disappear, the unit can sometimes be buried below the water table. Drawing the water by pumping instead of from an open hole has two big advantages: since the water is not exposed to sun and air, there is less loss by evaporation and the source does not become polluted.

1.5.2 Water catchment dams

A bank of sand often forms where the stream enters and this should take a unit satisfactorily but, if the dry season level drops too far, the unit must be taken up and reinstalled suitably. Wind and water movement should keep the bed surface clean but, if it becomes blocked, this is dealt with as advised under SITE CLEANING AND MAINTENANCE.

1.5.3 Sea shore

The unit should have at least 30 cm water over it at spring low tide. In very permeable sand on a level beach, a unit can sometimes be installed above LTM; it must be as deep as possible but, unless water has free access to it at all times, flow may be limited at low tide. Salinity must be monitored during site selection if water is wanted for aquaria, marine laboratories, etc., for areas may be diluted by freshwater run-off or beach springs.

In contrast to rivers and dams, sea movements are daily and predictable, and a known factor in the suction head. Tidal pattern may vary widely: rise and fall may be from under 2 m to 15 m (6-50 ft) and the tide may recede anything from a few metres to over 500 m.

1.6 Sites with Little or no Sand

Soft mud areas cannot be used, but where the bed is of rock or clay it may be possible to apply the following general method. Excavate or blast a hole. The shape need not be regular but for top quality water the surface area should be about 2 $m^2/m^3/h$, with depth of at least 60 cm. Usually a much lower ratio is possible. The bottom should be filled with coarse gravel or small stones to a depth of 20 cm to prevent the unit from sealing itself on the bottom and also to provide good lateral access for water. The unit is then placed in position and water allowed to enter, after which the sand is poured in until it forms a hump over the hole. If two units are needed they should be well spaced in a long trench. This variation clearly needs more work than a standard site, but it can be applied to such as the following.

1.6.1 River

Sometimes it is possible to work on the actual river bottom, perhaps after putting in a coffer dam. Or a site may be prepared alongside the stream, to make an artificial 'bay' when complete. It is essential to avoid reaches with massive scour in the rainy season.

1.6.2 Storage dams

These are sometimes lined to stop seepage and care must be taken not to cut through this lining. Two prepared holes may be needed, to draw water at various levels.

1.6.3 Foreshore

The intertidal zone is sometimes too compact for direct use but conditions vary so widely that only general suggestions can be made, and technical advice is needed, especially where large volumes are required.

On most shores the problem is to excavate holes near enough to LTM to allow continuous pumping. Where there is much beach run-off a long trench parallel to the shore line, plus the water in the trench may allow at least an hour's pumping while the tide is off it. Perhaps a narrow trench dug seawards can give water access to the trench even at low tide.

Where only moderate volumes are needed, the hole can be dimensioned to hold a reserve of, say, two hours' pumping over low tide The simplest and cheapest solution is an excavation well up the beach which is pumped for a few hours while actually covered by the tide. This reduces suction head and run to a minimum. The formation of a hollow must be avoided where sea wrack, etc., can be trapped and buried. Such work is seldom easy but it is well worth considering, especially on a shore where the tide recedes so far that LTM is beyond practical reach.

1.7 Pipe Size and Suction Head

Resistance (friction head) rises rapidly with both rate of flow and reduction in pipe size, as Table 1 shows clearly. Bends, valves, etc., add to this resistance (for details consult a textbook on Hydraulic Engineering). Forty millimeters (1.5 in) is unsuitable for over 5 m^3/h, while 50 mm (2 in) pipe can handle 18 m^3/h if the pump is close to the water and not more than about 3 m (10 ft) above it. For any flow above 20 m^3/h at least 75 mm (3 in) pipe is essential.

Failure to understand the importance of total suction head (i.e., friction head plus static head, or height of pump above water level) can be both expensive and disastrous. A pump can suck only from a limited depth; a maximum of ≃ 7 m (23 ft), and with most pumps the flow has already dropped severely before this is reached. This is reduced by 1 m for every 1 000 m altitude. The suction line should therefore be dimensioned to have no noticeable resistance at maximum flow.

This is not so critical on the delivery side but the figures in Table 1 apply equally and the higher cost of a larger pipeline is recovered in economy on pump size and power used. As diameter increases, the carrying capacity of a pipe rises much more steeply than cost

1.8 The Pipeline

Flexible armoured hose is needed from the unit at least as far as the highest water level. It can be expected to last for at least ten years. Semi-rigid PVC-type can be used from the river edge and also for the delivery line. For river sites, local conditions will decide whether the line should be protected from vandalism, heating up, etc., by burying or in some other way. In the sea it is better to bury the line up to 50 cm deep if the terrain allows, or else secure it to rocks and so reduce damage.

Table 1

Resistance (friction head losses) as function of pipe diameter and flow

Pipe diam. in	Pipe diam. mm	Gal/h	Litres/h	Resistance/100 ft (30 m) ft	Resistance/100 ft (30 m) m
1-1/2	40	1 000	4 500	5	1.50
		1 500	7 000	10	3.00
		2 000	9 000	15	4.50
2	50	1 000	4 500	1	0.30
		1 500	7 000	2	0.60
		2 000	9 000	4	1.20
		3 000	13 000	9	2.70
		4 000	18 000	15	4.50
		5 000	22 500	24	7.30
3	75	Below 3 000 gal (13 000 litres) negligible.			
		3 000	13 000	1	0.30
		4 000	18 000	2	0.60
		5 000	22 500	3	0.90

1.9 Pumps

Land-based pumps must be self-priming, i.e., be able to suck air from the suction line and pull water until there is a continuous column from unit to pump. The method of priming varies and instructions in the pump manual must be studied. However large the complex of units, each must be developed individually, using a petrol-driven pump of 3-4 hp able to pass 2.5 mm (0.1 in) solids. Machine pumps should not be considered for village supplies unless the power to run them, whether electricity or petrol, can be guaranteed. Once the system has been established, a Patay hand-pump is recommended for regular use; this easily yields 4 m^3/h, it is strong and reliable, and its only running cost is a new diaphragm every 400 h running. Wind-powered pumps are not suitable for development but could perhaps be used for filling reservoirs, etc., if a hand-pump is always kept in reserve. A simple reservoir holding 2-3 days' supply makes the whole system most useful.

In the sea a shore-based pump should be used when the static head allows, but it must be safely housed above HTM. In a marginal site a sunk pump chamber, thoroughly watertight, may reduce the static head by up to 1.5 m. Tidal pattern and beach shape sometimes make a submersible pump essential; the normal grill is replaced by a special fitting connected by flexible hose to the unit(s). The pump should be mounted firmly on a pile or similar support at a point just exposed at spring low tide for easy maintenance. Submersible pumps are powered only by electricity through heavy-armoured cable. It may be undesirable to bring this down a beach to which the public has access but some types can be well protected by running the cable down the rising main, emerging by a special duct at the pump. A small submersible pump is now available that can be housed directly on top of the unit.

1.10 Site Cleaning and Maintenance

In marine sites tidal and wave movements keep the surface clear and this is also true of most river units. If any blocking occurs, this will be in the top 1-3 cm, usually 1 cm, and this is most likely in the still water of storage dams. If reduced flow, not due to pump or other factors, suggests that there is surface blocking, the following progressively vigorous actions may be taken:

(a) stop the pump and rake over an area of about 5 m radius around the unit, working to a depth of about 5 cm. Then pump to waste while redeveloping as long as needed;

(b) skim off about 3 cm sand and replace with new sand;

(c) fork over the area lightly and then back-wash by moving suction hose to pump outlet and drawing water through spare hose. Somebody should stand on the unit until the pipes are changed back and the redevelopment starts;

(d) remove unit and install and develop in nearby site;

Options: (c) and (d) will be needed seldom if at all. In a system with several groups of units, provision for back-washing can be made with little modification to the plumbing. If the suction lines are long, this can be used to fill them with water and so reduce priming time when restarting.

A change in tidal pattern or badly sited breakwater may remove a metre or more of sand from the beach, though this is unlikely near or below LTM. If the unit becomes exposed, it must be installed afresh.

The contours of a river bed may change, making it necessary to find a new site, though intimate knowledge of the river, with a careful preliminary survey, should allow the selection of stable reaches Where scouring occurs, it may be enough to replace the sand and develop again, but the trouble will probably recur and it is usually better to move the unit.

A firmly embedded unit can be quickly freed by changing over suction and delivery hoses at the pump and blowing back, after letting some air into the line to the unit.

2. THE APPLICATION OF THE SWS SYSTEM TO FISH PONDS

Since all ponds have impermeable bottoms, whether natural or artificial, beds for housing SWS units must be introduced. The following procedure is suggested.

2.1 Medium

Most sand seems to be dug from river beds where it is constantly being deposited. Much contains a high proportion of particles below 1 mm but the quality varies and average particle size is larger when the flow is fast. Sand should preferably have not more than 10 percent below 1 mm with most between 1 mm and 4 mm. If the sand is fine and dirty, it can be quickly washed in a concrete mixer with water hose running, the best procedure being found by trial.

2.2 Siting

A corner of the pond is best. Its existing banks form two sides, the others being made by walls of roughly piled rocks or blocks (see Figures 2 and 3). If the area is close to the overflow to next pond, it has most water movement over the surface. Nearly all pond water is already dirty and it is essential to draw from a pond where the mud is not being stirred up by fish such as large carp, for this can add unnecessarily to the filter load. It is best to use a small pond without any fish or divide off a section of a large pond containing only small or surface fish.

2.3 Size

An area of not under 3 m x 3 m is suggested, with a final total bed depth of about 65 cm in centre, i.e., including SWS unit, and tapering to about 30 cm at edges (see Figures 2 and 3).

Good sand is available at very low cost in most parts and the biggest beds practicable should be made. These will allow maximum volume and also last longer between maintenance. Even a bed 5 m x 5 m and 75 cm deep would cost under Rp. 20 000[1/]. At 24 m^3/h, this is equivalent to a surface flow rate of 16 litres/m^2/min, which is within the flow for maximum quality slow sand filtration.

2.4 Construction

Although the work can be done while the pond is full, it is easier and quicker when a pond has been drained for cleaning. The operation is as follows, all figures being taken as approximate:

(a) clear away mud to leave hard bottom;

(b) complete sides by building rough walls of rocks, old bricks, etc., 30 cm high;

(c) to allow good access of water and keep the unit above clay bottom, two different methods are possible and the more convenient should be used:

(i) put down a circle (1 m diam.) of stones 5 cm diam. and depth of 10 cm;

(ii) put a large rock or block 10 cm high under each corner of unit (see Figures 2 and 3);

(d) pour sand to depth of 15 cm and put unit in position. Press down firmly but do not stand on it. If pond is empty, let the water start filling it;

(e) add gravel/sand to water (to stop forming air pockets) until centre has total depth of 65 cm sloping to 30 cm at edges. If a large stone of 15-20 cm thickness is placed on top of unit, it indicates depth of sand and holds it steady.

1/ This section of the report applies to conditions in Indonesia

2.5 Development

Thorough development is essential - it may take a whole day, with many stop/starts to reach maximum volume and top quality. If the sand contains much fine material and progress is slow, it will help to blow back briefly at low speed several times while somebody stands on the unit to hold it in position. "Briefly" means not more than 2 or 3 minutes after the water has reached the unit. Several times in this way will be more effective than a longer single period.

As explained elsewhere, in the high-ruling temperatures of this area, continuous running is essential to maintain the aerobic conditions required by a biological filter. Once development by petrol-driven pump is complete, this should be replaced by a 2 or 3 hp electric pump.

2.6 Water Quality

The raw water in the Sukabumi and Grobogan fish ponds (Indonesia) was more polluted with organic matter than in any previous tests. In both, there seemed to be complete removal of all algae and organic particles, leaving only faint milkiness caused by fine clay particles already discussed. These particles are probably too fine to be taken out by any filter. At the end of development, particles will be removed to nearly 2 µm, including *Lernaea* larvae.

After nearly seven days of pumping, the bed will become a biological filter and the following extra results are expected:

(a) reduction of ammonia by around 80% and of BOD and detergent by about 50%;

(b) removal of above 75% of all bacteria but this could be as high as 98%;

(c) marked reduction in DO level, since the oxygen has been used to oxidize polluting materials. This is simply restored by cascading, etc.

Note: It is difficult to give definite figures for quality, because

(a) conditions in these ponds vary widely, especially as regards algae, both free and surface, and

(b) the temperature is higher than in any other outdoor work we have done. The biological activity is, therefore, high and results may be better than suggested.

2.7 Maintenance

Since most material removed is probably organic, this should break down leaving few solids, but only experience will show how quickly the volume is reduced by blocking. Simple raking of the surface followed by redevelopment while pumping to waste should not take above two hours, and may take much less. It may prove useful to do this on a routine basis.

Conditions are very different from ponds in temperate zones. It is suggested that a system be established at one or two centres and fully observed for at least a month before it is used generally, but it should give much improved water for hatcheries and fry-rearing.

2.8 The Need to be Patient!

Full preparation of site and very complete development are essential, even if each may take a day or more. A source is being established for use during months or even years and there is no point in trying to do this in the shortest time. Once the system has been formed and development begun, it may be enough to leave a workman in charge, stopping and starting frequently, and gradually bringing the volume up to full flow.

Although it is possible only in hilly country, gravity abstraction from SWS unit is worth considering for providing clean water to breeding ponds at no cost for power, once the system is fully established. It can run by siphon, but it is better to bring a pipe through the pond wall and thus have a permanent head of water. In theory, a total head of 1 m, i.e., from water surface to outflow, can give a volume of about 30 m^3/h from a 50-mm pipe. In such condition, considerably less could be expected but the pipe should be fitted with a control valve and also a U-joint to keep the line full of water.

The screen well, which is ideal for working at depth, is not recommended for artificial beds or shallow formations, for the following reasons:

(a) it appears easy to insert but it is equally easy to disturb. In contrast, pumping holds the SWS unit down;

(b) any erosion or removal of sand can quickly expose the screen to raw water;

(c) most important, when the unit is installed and covered, all water must travel down through the sand to the depth of the unit and then up. In fact, most of it travels much farther, from several metres. There is no such minimum imposed by the well screen;

(d) the bed can be much more fully developed by the unit.

2.9 Note

The construction of artificial beds for obtaining improved water from fish ponds is described above, with the possibility of running by gravity, thus cutting out cost of power after original establishment. This method is being used in several countries for supplying water to isolated communities in hilly areas.

The unit is placed in a sand-filled collection area, either natural or artificial, and a pipe is then taken through the containing bank or wall to ensure a permanent head of water. In some cases, only a 0.5 in pipe feeding into simple storage and running to waste is needed. This can provide 24 m^3 daily from a sand bed of 2 m^2, and unless the source is grossly polluted, the water will be close to western potable standards. We believe that this possibility is worth notifying to authorities concerned with water in areas beyond the reach of piped supplies.

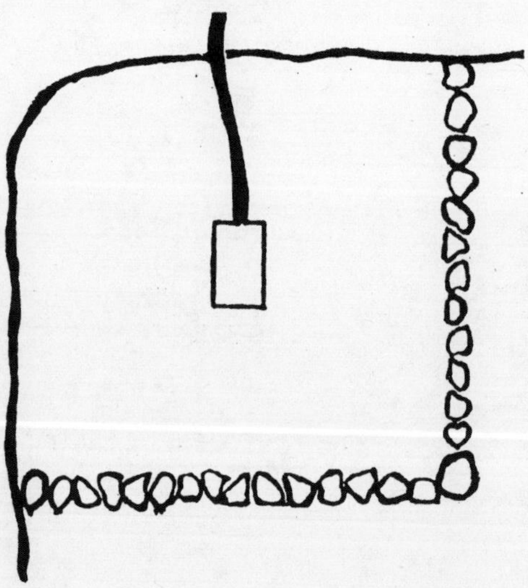

Figure 2. Plan (not to scale)

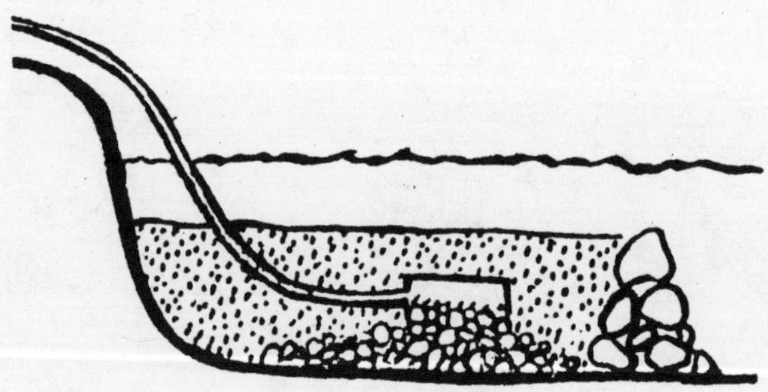

Figure 3. Section (not to scale)

FILTRATION FOR SALT OR FRESHWATER TANKS

Although the following note is mainly to meet the need for simpler and more effective filtration at the Brackishwater Centre, Jakarta, these two basic methods are effective for salt or freshwater tanks generally. Both are economical in space and material and are simply maintained. They are essentially biological filters and equally efficient, and are being used successfully in large tropical aquaria in UK with heavy stocking. The choice of type is by convenience, for one is likely to fit more easily into an existing circuit than the other. Designed for flows of up to 1 m^3/h, they can easily be scaled up by having two or more systems in parallel.

3.1 External Plastic Container

A dustbin (trash can) of 80-100 litre capacity is ideal, but if an old plastic drum of similar size can be found that is strong enough, this is equally good (Figure 4).

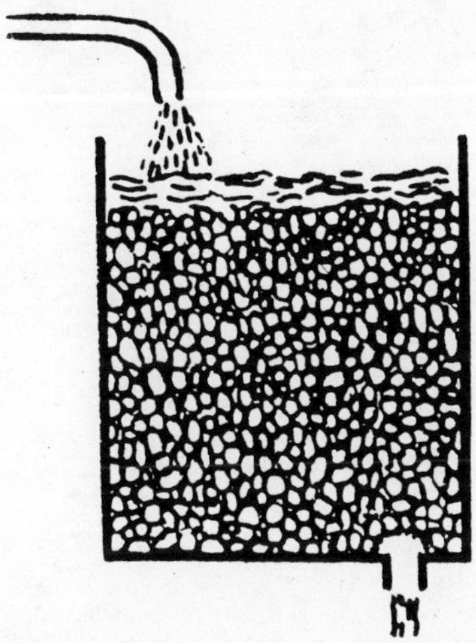

Figure 4. Drum unit for salt or freshwater tanks (not to scale)

Make a hole close to bottom and fit suitable outlet. Unless the container is to rest on a concrete slab, prepare a strong support for when full it may weigh above 200 kg. When in position, fill to within 15 cm of top with 2-5 mm gravel. Cover surface with sheet of glass or nylon wool in such a way that all effluent must flow through it. This sheet removes suspended matter and is changed or cleaned from time to time as it becomes visibly blocked. The mass of gravel becomes a biological filter within 7-10 days in a temperature of 26°-30°C and this should be left undisturbed as long as possible, certainly for at least six months. When this must be cleaned, some should be set aside, unwashed, to serve as culture to start biological filter quickly.

This method is most suitable when the existing main tank overflows to filter, storage, etc. Within reason, the larger the container the better, for it allows the water to flow over a larger active surface.

3.2 SWS Mini-Unit Housed in Gravel

This involves the use of a small electric pump - about 200 W - which may already be in use for circulation. An old container as discussed above is cut in half and filled with gravel in which the mini-unit is housed. The grade specified above should be used for the greater part, with a topping 4-5 cm deep, of clean, well washed sand (see diagram, Figure 5).

This is placed conveniently in the tank and serves as a self-contained sub-gravel filter. Massive development is not needed, but it should be stabilized by pumping to waste over 3-4 stop/starts for, say, 10 minutes. This biological filter takes at least a week to become effective, and it may help to add a little sand or debris from the tank bottom to seed it.

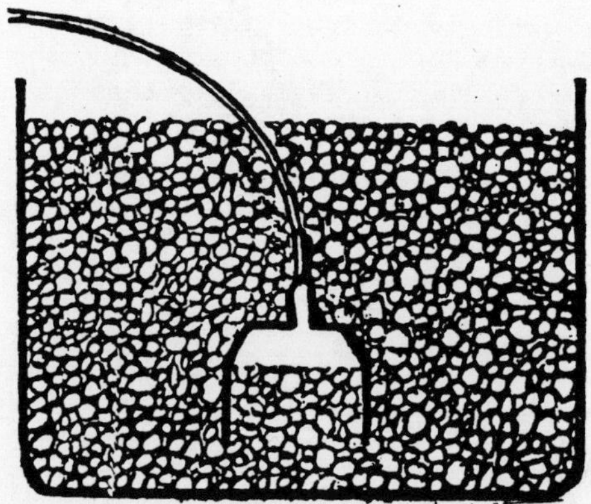

Fig. 5. SWS Mini-Unit Housed in Gravel
(Not to scale)

Standing clear of the tank bottom, it receives little silt: even in heavily loaded and fed tanks, the flow remains stable for 4-5 months without attention. The simplest way of cleaning, when reduced flow makes it necessary, is to remove the container, wash the sand/gravel and replace, also keeping a small amount of dirty gravel for seeding. This is suitable for systems housing small crustacean larvae, for it works by suction at a flow rate that does not draw down the larvae to the bottom surface. Accumulated fibrous debris on the tank bottom needs to be removed by syphoning from time to time.

Both of the above filters will process ammonia and other metabolites and reduce the need for water replacement, especially if the systems are only lightly stocked with larvae. It must be noted that this biological activity uses up some of the DO and this must be restored by cascading, injection, etc. One simple device for this is illustrated (Figure 6). Particularly with sea water, if there is any risk of a diminished pH, it is useful to use coral breakdown sand as a medium. This is easily obtained from many localities.

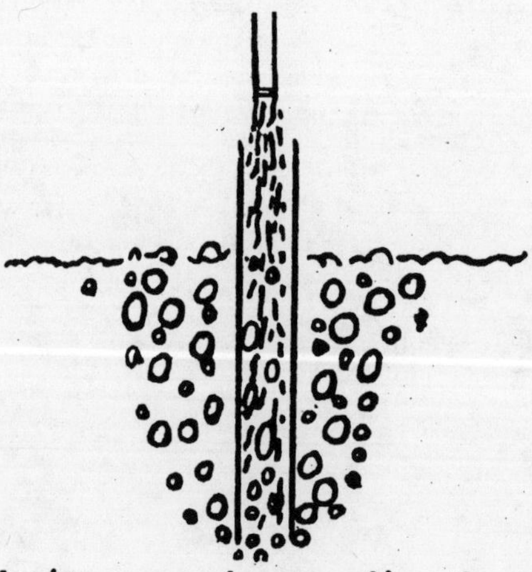

Fig. 6. Replacing oxygen by cascading the water.

APPENDIX 2

Upward Flow Filtration

Trickle sand bed filters, which are sometimes used in the treatment of hatchery water, block easily because the solids are retained on the surface of the bed. Frequent back-flushing is required to clean these filters.

Upward flow filters, such as is illustrated below, do not block so quickly and can be flushed without reversing the water flow by the use of increased water flow and air pressure.

The following notes are extracted from New *et al* (1974). The materials from which the filter is constructed can be altered and the size of the filter must depend on the demand for filtered water.

The filter consists of a 114-litre circular polyethylene filter vessel half filled with filtration media. Water enters the bottom of the tank and is distributed throughout the whole cross section of the filter media by means of a pipe grid constructed in 1.25 cm schedule 40 PVC. Several 0.5 cm diameter holes at 2.5-cm intervals are drilled in the underside of the grid pipes to distribute the water evenly. The 57 litres of filter media consist of a 10-cm layer of crushed rock (1.9-2.5 cm diameter) at the bottom, a 20-cm layer of pea gravel (0.3-0.6 cm diameter) in the middle and 5 cm of fine sand (0.05/m) topped with a 2.5-cm layer of 1.25-cm diameter dolomite or oyster shell. Filtered water leaves the filter by means of a 5-cm diameter PVC overflow pipe and is distributed by gravity.

A filter of this size copes adequately with a flow rate of 1 700 litre/hour and can be used as a biological filter[1/] in recirculation systems as well as a physical filter. When used as a biological filter, it should be seeded with bacteria by adding gravel from an old filter. Sea-water biological filters take longer to develop than freshwater filters.

Facility for flushing the filter with an air-water mixture is provided by a T-piece inserted in the line between the water pump and the filter, close to the latter. The air line is protected with a threaded ball non-return (check) valve, controlled by a PVC needle valve and supplied by a mobile compressor or diaphragm pump supplying at least 1 CFM at 20 psi. When the flow rate of the filter decreases due to gradual blocking, the outlet water is directed to waste while an air/water mixture is used to flush the filter. This causes the filter media to lift slightly and dislodge accumulated detritus.

Further reading on the topic of filtration can be found in Spotte (1970).

1/ If used as a biological filter, it has to be operated continuously, not intermittently

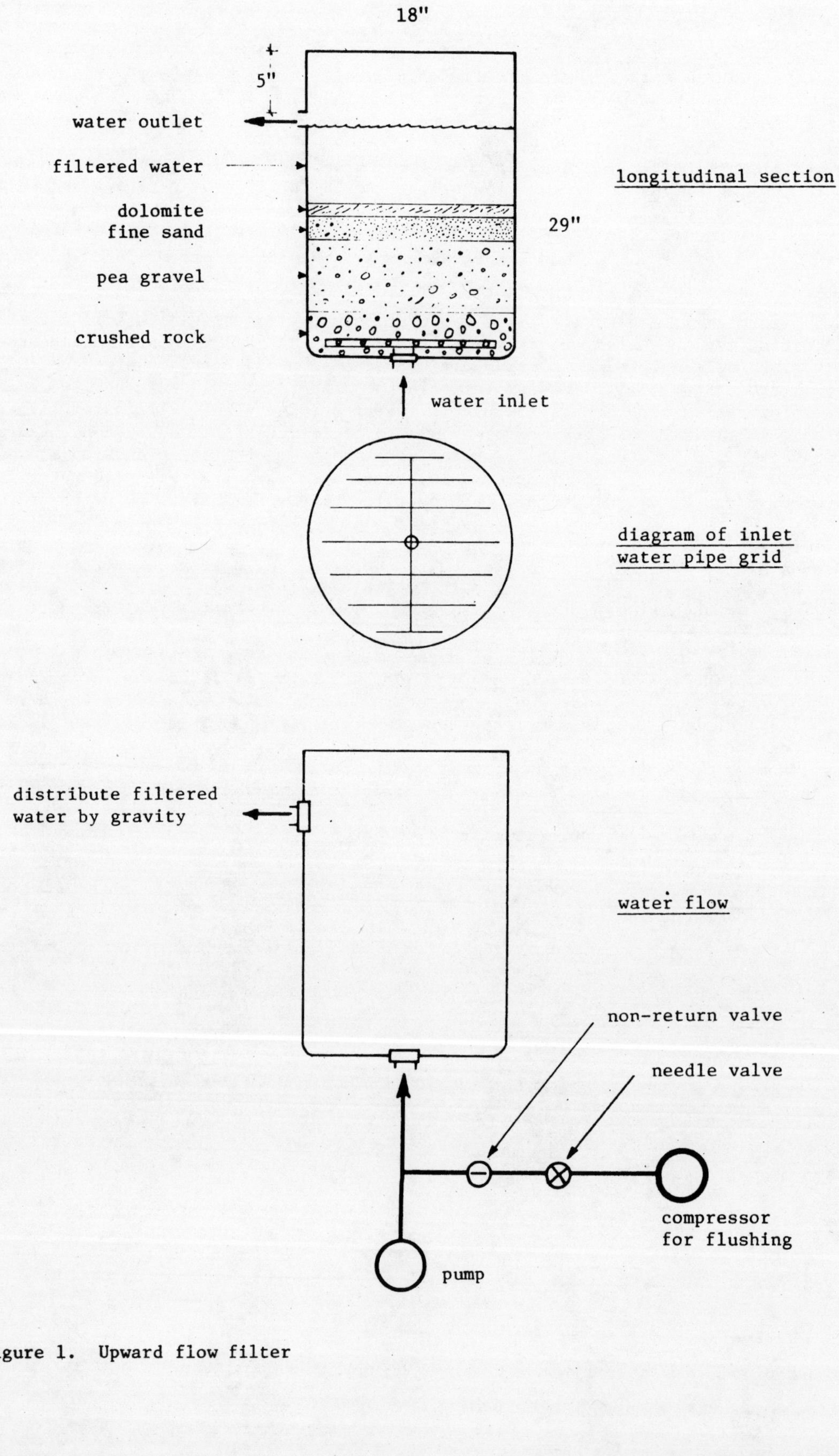

Figure 1. Upward flow filter

APPENDIX 3

Production of Brine Shrimp Nauplii (BSN) Feed for Larvae

1. CYST SOURCE

Artemia eggs (cysts) can be obtained from commercial companies operating from many countries including the USA, France, Spain, the Peoples Republic of China and Thailand. The quality of the cysts varies enormously from source to source and between individual batches from a single source (Vanhaecke and Sorgeloos, 1981). Methods of evaluating batches of *Artemia* cysts have been suggested by Sorgeloos (1978).

2. TREATMENT OF CYSTS

The hatchability of *Artemia* cysts can be increased by decapsulation. This, and an alternative and less severe version of the same treatment, also disinfect the brine shrimp eggs and remove or reduce an important source of disease from larval freshwater prawn rearing systems. Decapsulation also removes the need to separate the nauplii from the egg shells during the hatching process. The following alternative techniques for pretreating brine shrimp cysts before hatching are available, but require experimentation and modification for each source/batch of eggs if the best results are to be obtained.

3. DECAPSULATION

The two techniques, A and B, differ where indicated:

A. (Sorgeloos, 1978; Sorgeloos *et al.*, 1977); B. (Tunsutapanich, 1979)

(a) Hydrate 200 g of cysts in 3 litres of seawater or freshwater with vigorous aeration for one hour.

(b) Filter and wash the eggs on a 150-200 μm screen.

(c) A: Add 6 litres of decapsulation solution to the cysts (a mixture of 3 litres of commercial hypochlorite - 'Chlorox', 'Oldrox', 'Sanichlor', or similar in which 148 g of commercial sodium carbonate has been dissolved with 3 litres of water). Keep the suspension of eggs to 15°-20°C by adding ice, in tropical climates.

B: Add water and ice to a volume of 1.4 litres (maintaining the temperature at 20°C with ice) to which 25 g of technical calcium oxide and 55 g of commercial bleaching powder (Calcium hypochlorite) is added.

(d) Stir continuously and add more ice to keep temperature below 40°C.

(e) A: Nil.

B: After 5-8 minutes, add ice to reduce the temperature to 20°C and repeat the addition of the chemicals in (c) above.

(f) After the egg suspension becomes orange in colour (5-10 minutes), filter through a 150-200 μm mesh and thoroughly wash with tap water or seawater until there is no more smell of chlorine.

(g) Resuspend the cysts in 2 litres of seawater or freshwater containing 1 g of technical sodium thiosulphate.

(h) Stir well for 2-5 minutes.

(i) Allow decapsulated cysts to sink.

(j) Siphon off debris and unbleached eggs from the surface.

(k) Filter off the decapsulated cysts with a 150-200 μm mesh and wash well with water.

A. (Sorgeloos, 1978; Sorgeloos *et al.*, 1977)

(1) Either: use for hatching now
or: suspend the decapsulated cysts in a saturated brine solution (300 g technical NaCl/litre); aerate for 5 minutes; stop aeration (cysts float; debris sinks); drain or siphon off debris; aerate 2 hours; filter and resuspend in fresh brine; aerate for two hours; store in fresh brine for use later.

B. (Tunsutapanich, 1979)

Either: use for hatching now
or: add technical salt as a solid to the drained cysts in the ratio 30 g NaCl/100 g cysts; pour off the resultant liquid and store at room temperature.

(m) Do not expose decapsulated cysts to sunlight.

4. DISINFECTION

This treatment does not fully decapsulate the eggs but disinfects them thoroughly.

(a) Place 4 kg of eggs in a bucket and hydrate for about one hour with fresh or sea-water.

(b) Drain off the water, add 0.5 kg of commercial bleach (calcium hypochlorite) and stir without water.

(c) Keep below 40°C with ice and leave for 10 minutes.

(d) Place in a phytoplankton net and wash thoroughly with water.

(e) Place in a 0.05 percent solution of sodium thiosulphate for 2-5 minutes, remove the net and squeeze out the liquid.

(f) Wash thoroughly with water. The disinfected eggs are now ready for the hatching procedure.

5. HATCHING

A great deal has been written about techniques for hatching brine shrimp nauplii (BSN) from *Artemia* cysts. References to these works will be found in Sorgeloos (1978), Sorgeloos *et al.* (1977) and Vanhaecke (1981). Those seeking to maximize efficiency and reduce the cost of using *Artemia* are advised to read these papers. Many systems have been devised to separate nauplii from unhatched eggs and from discarded shells, where non-decapsulated cysts are used.

In practice, most large hatcheries do not separate the shells and unhatched eggs from the nauplii before feeding them to freshwater prawn larvae. The hatching procedure for *Artemia* cysts is therefore simple.

Virtually any type of container can be, and is, used for hatching *Artemia* including rectangular and round tanks, garbage cans, converted chemical carboys and 'klong' (water) pots. The latter type of containers have to be siphoned to empty them of BSN after hatching. The easiest to use is a conical based circular plastic or fibreglass tank which is elevated so that the BSN can be harvested by gravity.

A tank with a one-ton capacity, stocked with 250-1 000 g[1/] (0.25-1 g/litre) of *Artemia* cysts, will supply enough nauplii for five 10 m^3 larval rearing tanks for one day. A hatchery with five tanks of this size has a potential production of 0.5-1 million freshwater prawn post-larvae per cycle.

1/ These figures are approximate. Actual quantity of brine shrimp eggs used depends on their source (hatchability and nauplii size)

The preferred BSN hatching tank is shaped as follows:

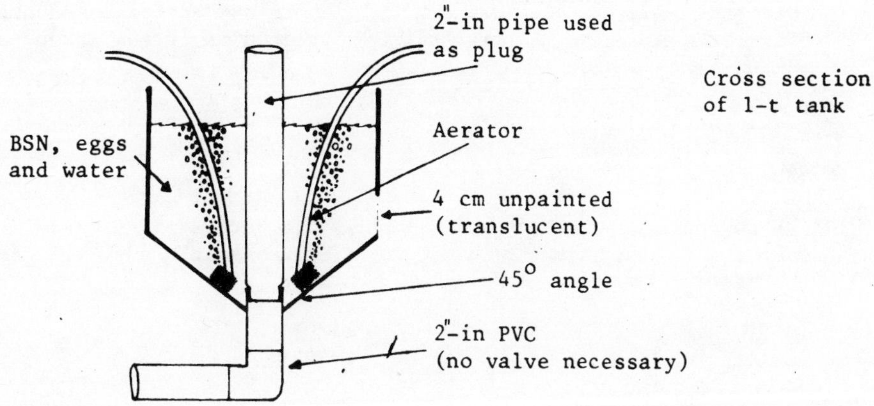

The hatching procedure is as follows:

(a) Insert the 2-in pipe plug in the drain of the circular, conical bottom 1-t tank and fill with water1/.

(b) Weigh in 250 g - 1 kg of *Artemia* eggs2/.

(c) Aerate vigorously.

(d) Maintain aerated at ambient temperature for 24-48 hours3/

(e) One hour before harvesting, add 50 ml of formalin (50 ppm) to the water4/.

(f) Cover the top of the tank and stop aeration, allowing the nauplii to gather at the bottom of the tank where the sides are translucent.

(g) Harvest through 250 µm phytoplankton net in a bucket5/ by removing the drain plug/pipe from the *Artemia* tank. The *Artemia* rearing water is discarded, not added to the prawn larval tanks. Unhatched eggs are also harvested and often hatch later in the larval tank. The drain plug is inserted again before the tank runs dry, thus retaining most of the egg shells, which float.

(h) (Optional) rinse the BSN with 12‰ salinity water.

(i) Remove the phytoplankton net from the bucket and divide the BSN between the prawn larval rearing tanks. Obviously, it is not necessary to drain the whole of the *Artemia* hatching tank at once: BSN can be drawn off as required.

(j) Start a new batch every day.

Notes: 1/ Many different hatching salinities for *Artemia* are suggested in the literature. They vary according to the cyst source and refinements to ease the separation of BSN and eggshells after hatching. We do not do this separation procedure and find that we get excellent hatchability using 12‰ water. The BSN experience no salinity shock when added to the prawn larval rearing tanks, which are also at 12‰. It is an important feature of the use of BSN, as a food, that it be a live feed.

2/ Amount depends on stage of prawn larval development. These weights are sufficient to supply five 10-t prawn rearing tanks with BSN (Manual Section 5.3).

3/ Feeding prawn larvae with 24 hours or 48 hours old BSN depends partly on the rate of *Artemia* hatching (characteristic of the source and the environmental conditions you supply it with) and partly on operator preference. We currently use 24-36 hours old BSN. Some hatcheries use 24-36 hours BSN at first and progress to the use of larger 48 hours or 72 hours old BSN (reared on 'green water' or rice bran) as the prawn larvae grow.

4/ This further disinfects the BSN eggshells and non-harvested eggs, especially reducing protozoa which would otherwise be added to the prawn larval tank with the BSN.

5/ Thus keeping the BSN in water during the harvesting process. The addition of an aerator to the bucket of this stage increases BSN survival rate, especially when harvesting is slow.

APPENDIX 4

Production of Prepared Feed (PF) for Larvae

1. INTRODUCTION

As mentioned in Manual Section 5.3, many differently prepared feeds are used in the rearing of freshwater prawn larvae in addition to the feeding of BSN. This Appendix describes the use of two types of PF.

The first PF is the one we currently use ourselves, an egg/mussel mixture, which we find is less problem to feed than the second PF, fish flesh. Fish flesh, especially when carelessly prepared or overfed, can be a grave source of water pollution in larval rearing. These two feeds can be prepared as follows.

2. PF 1: EGG/MUSSEL MIXTURE

(a) Blend 0.5 kg of shelled mussel (other molluscs can be used, but mussel seems best) in a blender.

(b) Strain the chopped mussel through a coarse cloth and discard the connective tissue, retaining only the material which passes the strainer.

(c) Using the whole of the mussel which has passed the strainer, add three or four whole eggs and stir thoroughly in the blender.

(d) Steam the mixture over water (like poaching an egg) until it solidifies into a custard.

(e) It may then be used for feeding directly (after screening to size - see Manual Section 5.3) or refrigerated for a few days for later use. Frozen material is not as good as fresh for feeding purposes.

3. PF 2: FISH FLESH

Skipjack tuna, bonito or mackerel are usually the fish of choice for preparing this feed. It may also be used as an ingredient in PF 1, partially or totally replacing mussel. The results with mussel seem superior.

(a) Fillet the fish, discarding head, bones and viscera.

(b) Grind and liquidize the flesh, as for mussel in PF 1, in a blender.

(c) Force the flesh through stainless steel sieves with a strong jet of fresh water. This grades the particles and washes the flesh free of blood and excess oil. The mesh sizes should be chosen to produce particles of a size relevant to the age of the prawn larvae (Manual Section 5.3).

(d) The fish flesh may be used directly or formed into balls of known weight for storage. It may be kept in the refrigerator for 2-3 days or frozen for longer periods. Frozen material is less satisfactory than fresh.

APPENDIX 5

Key to Larval Stages of Freshwater Prawns
Macrobrachium rosenbergii

The following is a simplified key to the eleven larval stages (Uno and Soo, 1969) of *M. rosenbergii* and is illustrated with some micro-photographs kindly supplied by Takuji Fujimura. The 'prominent characteristic' is some feature which appears for the first or only time at that stage:

Stage	Prominent characteristic[1/]
I	Sessile eyes
II	Stalked eyes
III	Uropods appear
IV	Two dorsal teeth on rostrum
V	Telson narrower and elongated
VI	Pleopod buds appear
VII	Pleopods biramous and bare
VIII	Pleopods with setae
IX	Endopods of pleopods with appendices internae
X	Three or four dorsal, teeth on rostrum
XI	Teeth on half of upper dorsal margin
POST-LARVA	Teeth on upper and lower margin of rostrum (also behavioural changes, mainly in swimming)

1/ Definition of the terms used will be found in Appendix 9

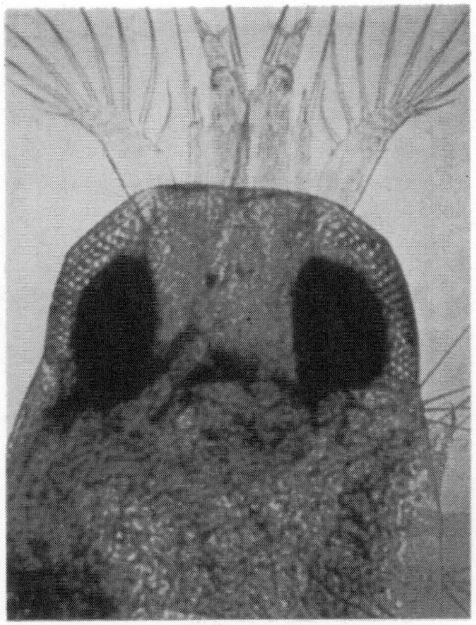

Stage I

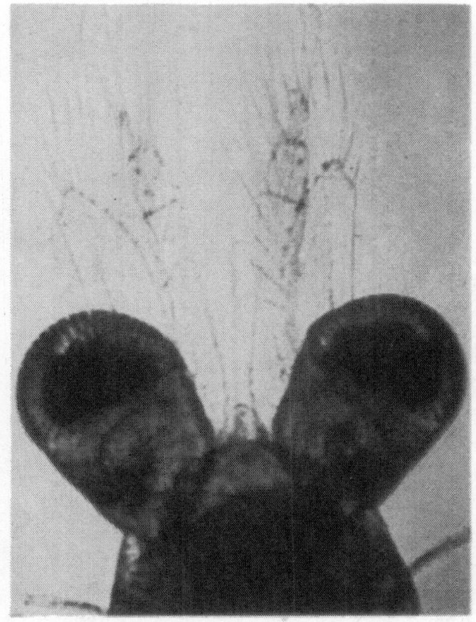

Stage II

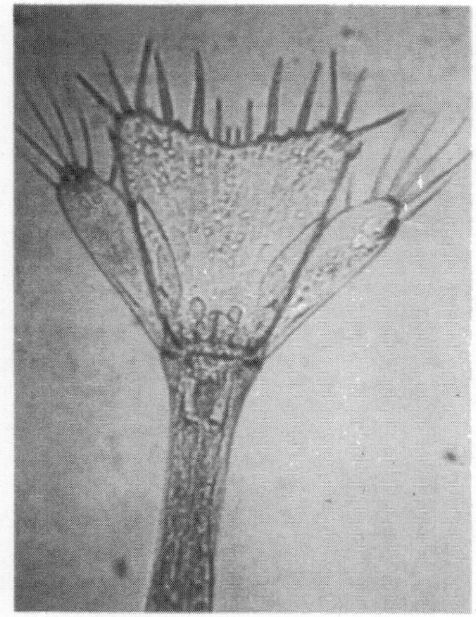

Stage III

Stage IV

Stage V

Stage VI

Figure 1. Key to larval stages of freshwater prawns, M. rosenbergii. Stages I - VI

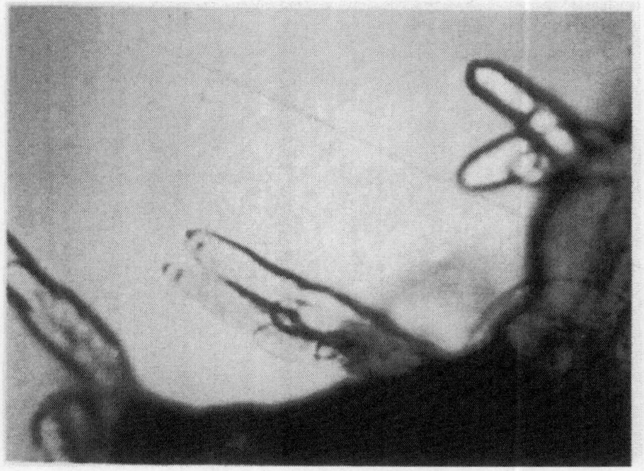

Stage VII

Stage VIII

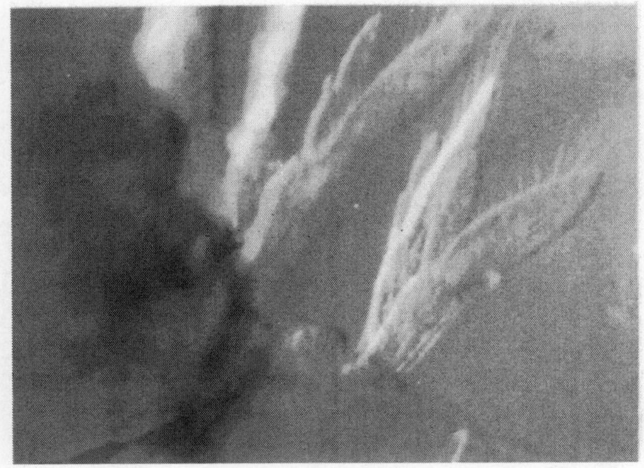

Stage IX

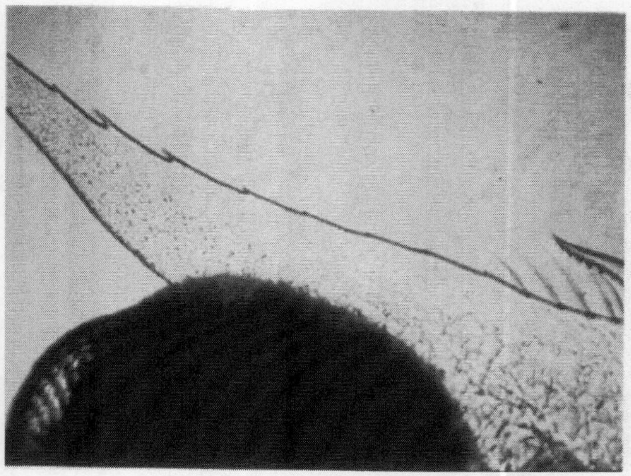

Stage X

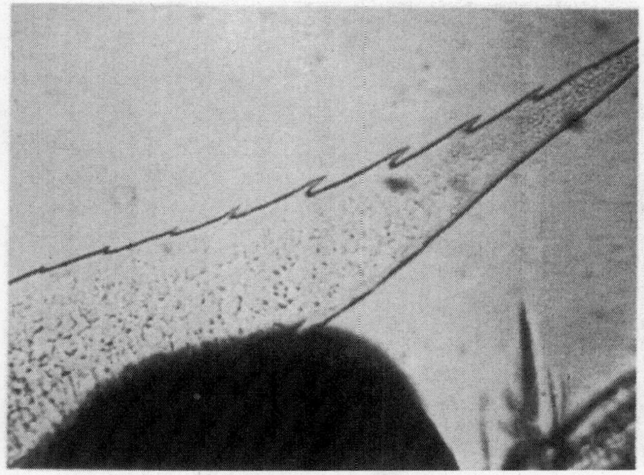

Stage XI

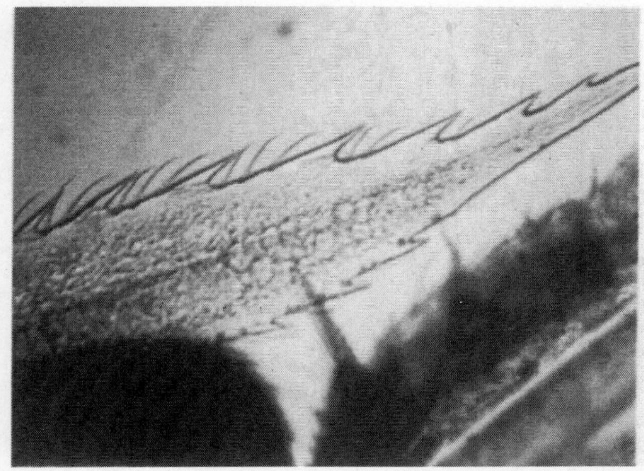

Stage XII

Figure 1 (continued). Stage VII - XII (postlarva)

APPENDIX 6

Stock Estimation

Estimation of the number of animals present under hatchery or pond conditions remains inaccurate, despite much thought and experimentation. The four critical times when it is important to assess the number (and sometimes the size) of prawns present in the system are:

A. When post-larvae are harvested. This provides a record and an assessment of the production efficiency of each larval batch and tank. Counting the larvae stocked into the larval tanks is not important (Manual Section 5.1).

B. When post-larvae are transferred from the hatchery to the pond. This is necessary to control stocking density and determine feeding rates.

C. At intervals during the growing period in the pond, to check growth rate and survival.

D. When prawns are cull-harvested or drain-harvested. This provides a final or cumulative record: an assessment of the productivity of the pond and management system being used.

The following methods are suggested for stock estimation.

A. At post-larval harvest

(a) Suspend the harvested post-larvae temporarily in a small and known volume of aerated water before transfer to the holding tank.

(b) Agitate the water thoroughly to evenly disperse the animals.

(c) Take two samples from the container in 100-ml beakers.

(d) Place the bulk of the post-larvae into the holding tank; do not wait until the sample counting process is complete.

(e) Count every animal in each of the two 100-ml beakers. A way of doing this is to take quantities into a graduated pipette hold at a 45° angle towards a lamp and to count the animals as they swim up towards the light.

(f) The number of post-larvae found in each 100-ml beaker is averaged and multiplied by the volume of water in the container mentioned in (a) above (in ml) and divided by 100.

Example: 90 and 100 post-larvae are counted in two 100-ml samples taken from a 25-litre container. The number of post-larvae in the 25-litre container is calculated as follows:

Average of 90 and 100 = 95; therefore, there are 95 pls in 100 ml of water, or 0.95/ml.

25 litres = 25 000 ml; therefore, there are 25 000 x 0.95 post-larvae 23 750 in the 25-litre container;

B. At post-larval transfer for stocking purposes

The following procedure, though ostensibly inaccurate, is sufficiently practical for use, especially where the same person always does the counting:

(a) On each occasion an initial batch is counted out individually by dipping a hand bowl into a larger bowl containing post-larvae. The batch individually counted may be 100 or 1 000, for example.

(b) Further batches are measured by visual comparison to the counted batch. The operator quickly becomes able to accurately estimate the number of post-larvae in each small bowlful dipped from the larger bowl.

(c) Once counted into batches, usually of 1 000 or 2 000, and placed in plastic transport bags at the hatchery, the post-larvae are not further counted at the pond site.

Notes: (i) Two-month old juveniles are easier to count than post-larvae. Manual Section 8.8.1 describes the use of nursery ponds.

(ii) Post-larvae can also be estimated by weighing the first (counted) batch and using weight alone to mesure subsequent batches.

C. During pond rearing

Once the prawns are in a pond, it is extremely difficult to estimate growth rate or survival. This topic was discussed at great length during a recent conference on freshwater prawn farming (New, 1982) and no satisfactory solution was presented. Multiple seine and cast net samples seem the only reasonable method of following the growth rate of a crop of prawns. At least a comparative estimate can be made. It is important that the method of sampling on each occasion is exactly the same (the same net; the same time of day; the same areas of the pond sampled; the same method of casting or pulling the net through the pond)

Barnes (1982) has used two 2 m x 1 m x 1 m cages in each pond and sampled these, subsequently using the results to predict what was happening in the rest of the pond. The cages were kept close to the bank, but at the average depth, and were resting on the mud. They were stocked at the same rate as the rest of the pond. Unfortunately, others have experienced poorer results from cages than the ponds they were placed in, probably due to the use of too fine a mesh, which blocks, causing poor water quality. Cages also provide a larger surface area for the prawns and thus a different environment than the ponds.

Even though the result may be grossly inaccurate, sampling animals by cast or seine net from the ponds on a regular basis does give the farmer some idea of how his crop is progressing (Figures 46 and 47) and is also a good opportunity to examine the health of the animals. It does not, unfortunately, give him more than a vague idea of survival rate.

D. At harvest time

From the farmer's rather than the scientist's point of view, there are two vitally important data which must be recorded at harvest time. One is the total drained weight of the harvest and the other is the average size of the animals harvested. From these two figures the numbers of animals harvested can be estimated. This, in turn, provides an indication of survival rate. Together, these data are used to assess the productivity of the pond and its management system. With the costs of production and the market value of the product, this information allows the economics of the pond to be determined.

Although the length of the prawn (biologists usually measure them from behind the eye stalk to the tip of the tail; farmers from the tip of the rostrum to the tip of the telson) is a more accurate form of measurement than weight, it is not so easy for the farmer to measure, particularly as he has then to convert the measurement to weight using a calibration curve (Figure 47). The weight of the animal can easily be measured on a portable scale. It will be inaccurate because of the amount of water adhering to the animal, particularly within the gill chambers. However, especially when the same person always does the weighing, it is possible to standardize the weighing technique and to achieve reasonable comparative accuracy.

It is suggested that about 250 animals are individually weighed for each 500 kg harvested (2 percent of the population if the average weight is 40 g). Take the subsample from the total harvest by dip-netting from the holding container or cage; selection of individual animals will inevitably lead to a bias toward the larger animals and is not recommended.

Typical length frequency distributions of freshwater prawn populations are given in Figure 46, while Figure 47 plots the length weight relationship of post-larval *M. rosenbergii*.

APPENDIX 7

Pond Feeds

Some examples of feed formulae for the pond rearing of freshwater prawns are given in this Appendix. The following general instructions are for the preparation of a diet which may be fed moist or dried.

1. Mix all the dry ingredients (except vitamin mix, if used) thoroughly, preferably by machine.
2. Add the vitamin mix, if used, and remix as in 1 above.
3. Add any liquid ingredient, such as fish oil or any wet materials such as chopped trash fish.
4. Remix thoroughly.
5. Add up to 30-35 percent of water. The quantity depends on the moisture of the ingredients, but sufficient water to produce a thick paste should be added.
6. Continue mixing. Passing the mixture through the coarse die of a chopper (see 7 below) does this job well and helps the binding characteristics of the diet.
7. Form into small plates, balls or discs by hand or, preferably, extrude through the 1/8-in diameter die holes of a meat chopper (Figure 43).
8. Feed as a moist diet on the same day or stir the extruded 'worms' to form 1-2 cm long pellets and sun-dry. 'Worms' dried on concrete in direct sunlight in Thailand are dry enough in six hours. A simple solar drier (Figure 44) can be constructed for drying pellets during the monsoon season. The pellet moisture should be reduced to less than 10 percent and the drying process must be as rapid as possible to prevent fungal growth. Dried pellets can be stored (cool and dry) up to 2-3 months.

The formulae given below are only examples; many other feeds have been used, or are possible, depending on the availability of raw materials.

DIET A (Balazs *et al.*, 1973)	Inclusion Rate (%)
Soybean meal (47% protein)	21.0
Fish meal (57% protein)	20.3
Shrimp meal (45% protein)	20.0
Ground corn	17.3
High gluten wheat flour (16% protein)1/	20.0
Iodized salt	0.4
Micro-ingredient mix2/	1.0
	100.0

Final ration is approximately 35% protein.

1/ Nutritional binder

2/ Consult original reference for constituents of micro-ingredient mix

DIET B (Balazs and Ross, 1976)	Inclusion Rate (%)
High gluten wheat flour[1/]	20.0
Iodized salt	0.4
Micro-ingredient mix[2/]	1.0
Ground corn	42.0
Soybean meal	12.6
Tuna meal	12.0
Shrimp meal	12.0
	100.0

Final ration is approximately 25% protein

DIET C and D (Boonyaratpalin and New, 1982)

	Diet C Inclusion Rate (%)	Diet D Inclusion Rate (%)
Fish oil[3/]	3.0	3.0
Shrimp meal	25.0	10.0
Fish meal	10.0	4.0
Peanut meal	5.0	2.0
Soybean meal	5.0	2.0
Broken rice	25.5	39.0
Rice bran	25.5	39.0
Guar gum[4/]	1.0	1.0
	100.0	100.0
Final ration is approximately (% protein)	25	15
Ingredient cost of diet (in Thailand in 1980) in US$/t	266	224

The following three diets are examples of commercially[5/] available feeds designed for marine shrimp which have been used in freshwater prawn culture. The actual quantities of each ingredient used are not declared by the manufacturers.

DIETS E and F (Ralston Purina literature, 1980)

DIET E: Purina Experimental Marine Ration, 20

Crude fat $\nless$ 5.0%
Crude protein $\nless$ 20.0%
Crude fibre $\ngtr$ 5.0%

DIET F: Purina Experimental Marine Ration, 25

Crude fat $\nless$ 10.0%
Crude protein $\nless$ 25.0%
Crude fibre $\ngtr$ 5.0%

1/ Nutritional binder

2/ Consult original reference for constituents of micro-ingredient mix

3/ From Sepat Siam (*Trichogaster pectoralis*)

4/ Binder

5/ The inclusion of branded products in this manual implies neither recommendation nor that substitutes are unavailable

The major ingredients of the diets, in addition to vitamin and trace mineral mixes, were:

DIET E

Fish meal
Soybean meal
Ground wheat
Wheat middlings
Meat and bone meal
Dried whey
Animal fat preserved with BHA

DIET F

Fish meal
Soybean meal
Meat and bone meal
Animal fat preserved with BHA
Ground yellow corn
Ground wheat
Wheat middlings
Dried whey
Soybean oil

DIET G (Central Soya literature, 1981)

Master mix shrimp feed 969.R

Crude fat	$<$	2.5%
Crude protein	$<$	25.0%
Crude fibre	$>$	5.0%

The major ingredients of this diet, in addition to vitamin and trace mineral mixes, were stated to be:

'Processed grain products'
'Animal protein products'
'Plant protein products'
'Animal fat preserved with BHA'

It is noted that the ingredient list provided by the manufacturer of this diet is less specific than that of diets E and F.

APPENDIX 8

Seine Net Design

(for 'continuous culture' system)

The construction of a seine net designed specifically for cull-harvesting freshwater prawns from ponds was detailed in Hanson and Goodwin (1977), from which the following notes are extracted, with acknowledgements.

In this net, 3/8-in polypropylene is used for the floater line and 1/2-in nylon for the sinker. Nylon is soft and follows the contours of the pond, while polypropylene is light and floats yet is stiff enough to minimize sagging. Sinker lines smaller than 1/2-in tend to sink into the mud. The ropes should be soaked for 12 hours and wet-stretched and dried to prevent twisting.

Monofilament netting is best. Two-inch stretched mesh, double knotted, 17-lb test netting is used which is 7 ft stretched, in depth. Sixty-pound test monofilament is used to fix the netting to the floater and sinker lines, using a 'double clove hitch' at every third eye. Net ends are reinforced with a heavy strand of nylon to prevent tearing.

Sufficient floaters are used to prevent line sag because otherwise prawns can crawl over the net. Two and one-half inch diameter, 2-1/2 in long cylindrical plastic floats spaced at 11-in intervals are ideal. Moulded U-shaped loads are favoured against commercial sinkers as they have a smaller cross-section. Leads cut to 37 mm from a 3-mm thick sheet, weighing about 60 g, hammered onto the sinkers every 11 in are suggested.

The seine is constructed with a bag similar to that of a beach seine except that the top is left open and the distance between the floaters is reduced to prevent prawn from escaping over the top. A bag with floor dimensions of 15 ft x 9 ft, tapering down to 4 ft, will hold 200 kg of live prawns.

APPENDIX 9

Example of Management Strategy for "Continuous" Culture Technique

In the first (English) edition of this Manual, the authors specifically omitted any details of the management of freshwater prawn ponds to be operated on a "continuous" basis and harvested by regular culling because there are obviously so many different ways in which this technique could be applied. We thought that to describe a particular management strategy might lead to its application without reference to the specific conditions of each site, with possibly sub-optimal results. However, there have been several requests by readers of the Manual to provide an example of how such an intensive system of culture might be operated. An example is therefore now given in this appendix but we must stress and reiterate the importance of treating it only as an example. The strategy illustrated here must be modified to the particular circumstances of each individual farm. It is an example of an intensive system. Stocking rates are therefore quite high ($15/m^2/yr$) and the achievement of a successful yield from ponds operated in this way depends on many factors which have been described in this Manual including the dependability of the water supply, the control of predation, the matching of feed quantity to biomass levels and water quality, etc.

Table 1 presents an example of how a single 1 ha freshwater prawn pond could be managed using the continuous culling technique. The numbers of animals stocked, the stocking interval and the anticipated numbers of animals which will be harvested are shown. The assumptions made in constructing this model are shown.

In constructing the model shown in Table 1 we have assumed that peak monthly harvests occur about six months after the first culling of each stocking. Our example has been based on the second and subsequent stockings being 50 percent of the number in the first stocking. First re-stocking occurs in the 10th month, with subsequent restockings being made every six months. A further assumption made was that a total of 50 percent of the prawns stocked would eventually achieve the desired market size chosen (45 g +); the rest were assumed either not to have survived or never to have become large enough to be retained by the culling seine.

The management model illustrated in Table 1 has been based on the assumption that, at average water temperatures of 24-30°C and cull-harvesting of prawns of 45 g and above on a monthly basis, commencing at the 7th month after stocking, it would take 18 months from first stocking to completely harvest all the surviving prawns from that batch. This assumption is based on some of the original observations of Fujimura in Hawaii (Fujimura, pers.comm.) who suggested that ponds should first be restocked in the 13th month (rather than the 10th month we suggest) with half the number of prawns used in the original stocking and thereafter stocked every 6 months at the lower stocking density for continuous operation. He also suggested that the pond should be drain harvested, so that it could be cleaned and treated and repaired, every 3 years or so, or as soon as a growth slow-down was experienced.

Again, it is stressed that actual management procedure must be based on knowledge of the specific environmental and other conditions at the farm site. The farm manager must be flexible and ready to modify his re-stocking and harvesting policy based on availability of PL's, harvest results, growth rate, etc.

Even though the actual distribution of the # harvested each month from a particular stocking may be different to that portrayed in this example, if the distribution of each population is similar, the total number of prawns harvested each month would be reasonably constant from month 22 onwards. If the market demand is not seasonal this regularity of supply is desirable. This example covers one pond only; in practice the farm manager would have several or many ponds and could adapt his stocking strategy to produce a steady or a seasonally oriented supply of marketable prawns from the farm, according to the characteristics of its market.

The example given in Table 1 is based on re-stocking on a six-monthly basis. Providing the animals and labour are available, re-stocking (or proportionately fewer animals) could be done more frequently. More frequent stocking would be more reliable since it would mean that less reliance was placed on the performance of each individual batch of animals. Clearly the partial or complete loss of a single monthly stocking of 12 500 juveniles, through predation or poor survival resulting from unsatisfactory larval production, would be less serious than the similar loss of a 6 monthly stocking of 75 000 animals.

Table 1

A theoretical model for stocking and harvesting a 1 ha freshwater prawn pond (continuous culling technique)

Month	FIRST STOCKING		SECOND STOCKING		THIRD STOCKING		FOURTH STOCKING		FIFTH STOCKING		TOTAL # HARVESTED
	Number Stocked	Number Harvested	Number Stocked	Number Harvested	Number Stocked	Number Harvested	Number Stocked	Number Harvested	Number Stocked	Number Harvested	
1	150000	-									---
2		-									---
3		-									---
4		-									---
5		-									---
6		-									---
7		4000									4000
8		5000									5000
9		6000									6000
10		7000	75000	-							7000
11		7500		-							7500
12		8000		-							8000
13		8000		-							8000
14		7500		-							7500
15		7000									7000
16		6000		2000	75000	-					8000
17		5000		2500		-					7500
18		4000		3000		-					7000
19				3500		-					3500
20				3750		-					3750
21				4000		-					4000
22				4000		2000	75000	-			6000
23				3750		2500		-			6250
24				3500		3000		-			6500
25				3000		3500		-			6500
26				2500		3750		-			6250
27				2000		4000		-			6000
28						4000		2000	75000	-	6000
29						3750		2500		-	6250
30						3500		3000		-	6500
31						3000		3500		-	6500
32						2500		3750		-	6250
33						2000		4000			6000
34								4000		2000	6000
35								3750		2500	6250
36								3500		3000	6500
								etc		etc	etc
Total number harvested from each stocking		75000		37500		37500		(37500)		(37500)	
										Annual number harvested YR I	37500
										YR II	75000
										YR III	75000

Summary of assumptions for Table 1

- Harvesting starts in 7th month after each stocking and is complete after 12 months.
- ~5 g - harvest size; 50% of prawns stocked are harvested.
- Initial stocking at $15/m^2$; subsequent stocking at $7.5/m^2$ every 6 months. Average # of PL's required per year = 150 000.
- Projected yields YR I = 1 700 kg; YR II onwards = 3 375 kg.
- Stable environmental conditions (i.e., no seasonability of growth rate).
- Anticipated growth rate based on water temperatures of 24–30°C.

APPENDIX 10

Glossary of Terms and Conversions

SECTION A: TERMS

Section A of this glossary defines unfamiliar terms used in this manual. The definitions are intended to make the terms understandable to the novice rather than to the biologist. Additionally, the definitions given to the words apply specifically to freshwater prawns, not necessarily to other organisms.

Artemia	Synonymous with brine shrimp.
Bank	The elevated rim of a pond. Also called bund or berm.
Batch culture	A system of rearing prawns, involving the total harvest, by seining or draining or both, at a certain interval after stocking. The ponds are then drained before re-stocking (see continuous culture).
Berm	See Bank.
Berried	Egg carrying.
Brine shrimp	A small crustacean whose larvae are used to feed larval prawns.
Brood chamber	An area formed beneath the abdomen of the mature female by the expansion of the pleura, in which the fertilized eggs are carried before hatching. In this area the eggs are oxygenated by movement of the pleopods.
Bund	See Bank.
Clearwater	Larval-rearing water which does not contain green algae.
Continuous culture	A system of rearing prawns in ponds which involves continuous pond operation. Ponds are not drained for harvesting, nor completely harvested. The larger animals are regularly removed by seine net for marketing, leaving the smaller ones to grow on. The ponds are regularly restocked with post-larvae or juveniles (see batch culture)
Count	This term, used by shrimp buyers, refers to the number of prawns or prawn tails per unit weight. When using this term, it is important to state whether shell-on/head-on, shell-on tails or peeled tails are being described.
Crustacea	Group of animals including prawns, lobsters and crabs.
Decapsulation	The removal of the hard outer layer (shell) of *Artemia* cysts.
DO_2	Dissolved oxygen content (of water). Sometimes reported as ppm and sometimes as percent of saturation level.
Dorsal	Upper.
Endopod	Anatomical term. See Figure 1 for positions.

Food Conversion Efficiency (FCE)

(See also food conversion ratio)

The amount of food necessary to produce one unit weight (wet) of prawns. For example, if a pond produces 1 250 kg of prawns and 3 200 kg of food were used during the rearing period, the FCE is:

$$\text{FCE} = \frac{3\ 200}{1\ 250} = 2.56$$

It follows, therefore, that the lower the FCE is, the better the efficiency (of conversion into final product) of the food is. The FCE of wet feeds will be much higher than that of dry feeds because of the difference in moisture content. To directly compare two feeds with different moisture contents, it is necessary to convert the different food conversion efficiencies to a standard moisture content or to bring the relative cost of the feeds into consideration. The farmer will find the latter option more meaningful.

Example: Feed A has an FCE of 2.8 and a cost of US$ 482/t. Feed B has an FCE of 6.9 and a cost of US$ 215/t. Which is the 'better' feed from the farmers' point of view?

To produce one ton of prawn using Feed A would cost US$ 492 x 2.8 = US$ 1 377.60; using Feed B it would cost US$ 215 x 6.9 = US$ 1 483.50.

Feed A is therefore cheaper to use, although its unit price is more than twice than that of Feed B.

Food Conversion Ratio

As for food conversion efficiency, except that it is written as a ratio, i.e., a food conversion efficiency of 2.8 is written as a food conversion ratio of 2.8:1. This means 2.8 kg of food is necessary to produce 1 kg of prawns (live weight).

The two terms are frequently used synonymously. For example, you may often see an expression such as "the FCE of the diet was 2.8:1".

Genital pores

The openings of the reproductive organs to the exterior of the animal. In males they are between the fifth pair of walking legs and in females between the third pair of walking legs.

Gill chamber

The area at the sides of the 'head' of the prawn that contain the gills through which the prawn takes oxygen from the water and releases carbon dioxide during respiration.

Green water

Larval-rearing water with an induced density of green planktonic algae (Manual Section 5.8.2).

Head

The front 'half' of the animal, consisting of the head and carapace area.

H_2S

Hydrogen sulphide.

Juvenile

Animals older than 60 days after metamorphosis, but which are not yet mature. In their natural habitat, freshwater prawns at this stage can move against strong currents, climb rapids, and move across wet areas to other waters. They are very hardy by this time.

Larva

Singular of larvae.

Larvae

Animals during the first 16-30 days after hatching from the egg. They require brackish water and swim upside down, tail up and backwards. Form is also different from adults.

Metamorphosis — The process of transformation by which a larva becomes a post-larva and takes on the miniature appearance and the behaviour of an adult.

Moult — To cast the shell.

Ovigerous — Having ripe ovaries (refers to females).

Pleopod — Anatomical term. See Figure 1 for positions (see swimmeret).

Pleura — Anatomical term. See Figure 1 for positions.

Post-larva — Singular of post-larvae.

Post-larvae — A term usually applied to animals from the time of metamorphosis up to about 60 days later. This term and the word 'juvenile' are applied very loosely and sometimes synonymously Post-larval freshwater prawns swim and behave like adult prawns and, as they age, cling to or crawl on surfaces rather than swim freely in the body of the water.

ppm — Parts per million. A unit of chemical measurement used for reporting the levels of trace materials (e.g., oxygen dissolved in water) or of an additive (e.g., formalin or chlorox). It is equivalent to 1 ml/m^3, 1 g/t, 1 µg/g or 1 mg/litre, for example. Where this manual prescribes the addition of a substance at a certain level (say, 40 ppm), the actual amount to add can be calculated as follows:

Example 1 (measuring a liquid)

The manual says "add 50 ppm formalin". The volume of water being dosed is (for example) 250 litres. The expression 50 ppm (parts per million) means 50 parts of formalin to every 1 million parts of water (e.g., 50 ml formalin in 1 million ml water). As 250 litres = 250 000 ml, thus the amount to add (in ml) is: $\frac{50 \times 250\ 000}{1 \text{ million}} = 12.5$ ml

Thus, 12.5 ml of formalin must be added to 250 litres of water to provide a concentration of 50 ppm.

Example 2 (measuring a solid)

The manual says "add 0.4 ppm of copper sulphate". The volume of water that has to be treated is (for example) 1.5 m^3. Then the amount of copper sulphate to add should be:

$$1.5 \times 1\ 000 \times 1\ 000 \times \frac{0.4}{1 \text{ million}} = 0.6 \text{ g}$$ 1/.

PPT (‰) — Parts per thousand. A unit of measurement usually applied to salinity. Also expressed as ‰. Twelve ‰ water, therefore, means brackish water with a salinity of 12 ppt. The salinity of full sea water varies but is normally around 35‰.

Prawn — Synonymous with shrimp.

Rostrum — Anatomical term (see Figure 1 for positions).

1/ In these calculations it is assumed that 1 ml of water weighs 1 g; 1.5 m^3 of water, therefore, weighs 1.5 x 1 000 x 1 000 = 1 500 000 g

Sessile	Not on stalks (applies to larval eyes in the first stage).
Shrimp	Synonymous with prawn.
Supernatant	The clear liquid after a precipitate has settled.
Swimmerets	Synonymous with pleopods; the 'swimming' legs beneath the abdomen.
Tails	The abdomen, or rear 'half' of the animal.
Telson	Anatomical term (see Figure 1 for positions).
Uropod	Anatomical term (see Figure 1 for positions)
Ventral	Lower.
Walking legs	Anatomical term (see Figure 1 for positions).

SECTION B: ABBREVIATIONS

$<$	Less than.
$>$	Greater than.
$\nless$	Not less than.
$\ngtr$	Not more than.
US	United States (see this Appendix, Section C: weights).
N.A.	Not Analysed.
μm	Micron (sometimes written simply as μ).
mm	Millimetre.
cm	Centimetre.
m	Metre.
in	Inch.
ft	Foot.
yd	Yard.
m^2	Square metre.
ft^2	Square foot.
yd^2	Square yard.
ha	Hectare.
km^2	Square kilometre.
mi^2	Square mile.
ml	Millilitre.
l	litre.

m^3	Cubic metre.
ft^3	Cubic foot.
yd^3	Cubic yard.
g	Gramme.
kg	Kilogramme.
oz	Ounce.
lb	Pound.
cwt	Hundred weight.
psi	Pounds per square inch.
GPM	(Imperial) Gallons per minute.
MGD	Million (Imperial) Gallons per day.
CFM	Cubic feet per minute.

SECTION C: CONVERSIONS

Section C should be used in conjunction with Section B, which explains the abbreviations used. Please note that the words gallon and ton have different values depending on whether the source of the text you are reading is "British" or "American" in origin.

LENGTH:

1 µm = 0.001 mm

1 mm = 0.0394 in = 0.001 m

1 cm = 0.394 in = 10 mm = 0.01 m
1 m = 3.28 ft = 1.094 yd

1 in = 25.38 mm = 2.54 cm

1 ft = 0.305 m = 12 in

1 yd = 0.915 m = 3 ft

WEIGHT:

1 g = 0.0353 oz

1 kg = 1 000 g = 2.205 lb

50 kg = 110.25 lb

1 000 kg = 1 t

1 t = 0.9842 UK t = 1.102 US t

WEIGHT (cont.)	1 oz = 28.349 g
	1 lb = 16 oz = 453.59 g
	1 cwt[a/] = 112 lb = 50.80 kg
	1 US cwt = 100 lb = 45.36 kg
	1 t[a/] = 20 cwt[a/] = 2 240 lb
	1 US t = 20 cwt = 2 000 lb
	1 UK t = 1.016 t = 1.12 US t
VOLUME:	1 litre = 1 000 ml = 0.220 gallon[a/] = 0.264 US gallon
	1 m^3 = 35.315 ft^3 = 1.308 yd^3
	1 m^3 = 1 000 litres = 219.97 gallons[a/] = 264.16 US gallon
	1 ft^3 = 0.02832 m^3 = 6.229 gallons[a/] = 28.316 litres
	1 gallon[a/] = 4.546 litres = 1.2009 US gallon
	1 US gallon = 3.785 litres = 0.833 gallon[a/]
	1 MGD[a/] = 694.44 GPM[a/] = 3.157 m^3/minute = 3 157 litres/minute
AREA:	1 m^2 = 10.764 ft^2 = 1.196 yd^2
	1 ha = 10 000 m^2 = 2.471 acres
	1 km^2 = 100 ha = 0.386 mi^2
	1 ft^2 = 0.0929 m^2
	1 yd^2 = 0.836 m^2
	1 acre = 0.405 ha
	1 mi^2 = 640 acres = 2.59 km^2
TEMPERATURE:	$^oF = (\frac{9}{5} \times {^oC}) + 32$
	$^oC = \frac{5}{9} ({^oF} - 32)$
PRESSURE:	1 psi = 70.307 g/cm^2

---

a/ "British" or "Imperial" units